# CHESAPEAKE
# BAY
# COUNTRY

# CHESAPEAKE BAY COUNTRY

## Nature Notes & Country Cooking

### W. C. (Bill) Snyder
### *Illustrated by* Ravay Snow

design by
Patrick S. Smith

The Donning Company/Publishers
Norfolk/Virginia Beach

The Donning Company/Publishers
5659 Virginia Beach Boulevard
Norfolk, Virginia 23502

Edited by Tony Lillis

**Library of Congress Cataloging-in-Publication Data**

Snyder, W. C. (William C.), 1914-
  Chesapeake Bay country.

  1. Seasons—Virginia—Tidewater Region. 2. Natural
history—Virginia—Tidewater Region. 3. Cookery, American.
4. Cookery—Virginia—Tidewater Region.
I. Title.
QH105.V8S69      1987      509.755'1      87-21406
ISBN 0-89865-568-4 (pbk.)

**Printed in the United States of America**

*Dedicated to
My Family*

# Foreword

From home base on his beloved creek in James City County, William "Bill" Snyder observes the changes each season brings to Tidewater, Virginia. In his latest book. *Chesapeake Bay Country*, he chronicles those evolutions in four seasonal sections.

We go along on a snowy winter afternoon spent before a roaring fire in a tumbledown cabin in the woods. We're amused—and sympathetic—when he vividly describes his annual battle with deer flies. He takes us on a trip to Tangier Island in the middle of the bay. We go with him on fishing jaunts and berry-picking expeditions.

Snyder's writings are the result of an intimate lifelong interaction with wildlife and its habitats, and their relationships with man. He's learned to revel in nature's infinite variety, while respecting her power.

In one chapter, he comments, "Too many times, the beauty of the commonplace is unseen by the beholder." He is an eloquent spokesman for that world most of us look at but don't really see. *Chesapeake Bay Country* will give pleasure and new knowledge to nature lovers and casual readers alike.

In the book, he has combined nature notes with country cooking recipes. Each of the four sections ends with directions for preparing longtime favorite regional dishes. Many use the Chesapeake Bay's fruits of land and water. They're an added bonus. The real meat of the book is the telling of simple, everyday vignettes with awareness, humor, and sensitivity.

Hope Reece

# Introduction

Chesapeake Bay Country will never be covered in one volume or even many volumes. It is too good, immense, and complex for any writer to do complete justice. This book covers primarily a minute portion of that country located at Jamestown, Virginia, where Powhatan Creek flows into the mighty James River that in turn empties into Chesapeake Bay. In sight of my house stands a 450-year-old cypress tree that possibly sheltered the Powhatan Indians or the seventeenth-century Jamestown settlers. The area is so steeped in history that a night watcher sitting quietly on the creek bank can almost feel the ghosts of those gone beyond. The creek, marsh, and woodland are representative of most of Chesapeake Bay Country. Read and enjoy.

W. C. (Bill) Snyder

# Contents

# SPRING

# The Creek

*Far back, through creeks and inlets making,*
*Comes silent flooding in, the main.*

—Arthur Hugh Clough

Powhatan Creek is mentioned many times in this book, so first off I want to say a little about the "creek."

It is located in James City County, Virginia, and crosses Highway 31 about one and a half miles east of the Jamestown Ferry.

The creek was named for the Indian chief who won fame as the architect of the Powhatan Confederacy of Virginia, which at one time numbered thirty tribes, with a membership of about nine thousand Indians. He real name was Wahunsonocock, but he preferred to be called Powhatan after his favorite village, that was located on the James River between Jamestown and Richmond. He was also the father of Pocahontas, who allegedly saved the life of John Smith, a leader of the first colonists.

The headwaters of the creek are born from springs north of Jamestown, and it winds in sinuous fashion to the James River nearly ten miles away. The creek can be navigated for over a mile up into the woods.

Powhatan Creek is of many moods, varying with the weather. Northeast winds push the tide far in, covering the mud, logs, and all plant growth except the few tall reeds and wild oats that are left from bygone days. When these abnormal tides occur in the fall, sora hunters pole flat-bottom boats over the marsh, and shotguns bang away at this small member of the rail family.

Conversely, when the winds move around to any compass point out of the west, they sweep the waters out of the creek like a giant broom. At this time, the boatman needs to know the channel to keep from becoming a part of some mud flat until the next high tide. This wind is favored by muskrat trappers in the winter because so much of the marsh can be covered.

Powhatan Creek was not like this in the early days. Then a deep channel made up from the river, and I am told that sailing vessels and lighters came into the Creek to pick up and discharge cargoes at various wharfs along the way. My home is located at one of the old landing places, and slabs of wood still jut out into the water and afford excellent hiding places for bass, bream, and crappie. A large pine stump still has a ring imbedded where boats tied up, and it is easy to shut one's eyes and imagine that ghostly voices of boatmen, lumbermen, and roustabouts speak to each other over the moving water.

There is much wildlife in the creek, and the surrounding woods and marsh. Low tide leaves mud flats where tracks of 'coon, muskrat, 'possum, mink, and an occasional otter can be seen. They travel the waterways constantly, feeding on crawfish, minnows, frogs, mussels, and other inhabitants of the swampy terrain. The crawfish are sometimes dug out of their inverted gray-colored mud houses that resemble small nests of hornets. When I was a boy, a favorite sport was dropping a string or long twig down one of the holes, and when the crawfish grabbed hold with a claw, it would be yanked out quickly. They are capable of quite a hard pinch with either of the claws. Boiled like lobsters, the little fellows are considered quite a delicacy.

The creek and its environs are home to many kinds of birds. Great blue and little green herons are plentiful, as are several species of woodpeckers. In the winter, robins abound and feed on berries of the holly trees that grow around the swamp. An occasional woodcock, with its long probing beak, is flushed once in a while, and the honeysuckle patches along the edge shelter coveys of bobwhite quail. Hawks, owls, and wood ducks use the hollows in dead stubs and tall hardwoods for nesting sites.

I walked a section of the swamp last year and was amazed at the size of some of the trees that were lumbered in the past. I did not have a tape measure, but some of the stumps were over five feet in diameter. One gigantic oak, dead for many years, supported the growth of a tree fern some twenty-five feet above the ground. I looked at another on a maple only five feet above the ground. In nature's mysterious way, the tiny spore had lodged, taken root, and turned into a small plant of beauty. It had six fronds ranging from one to three inches in length and they were emerald green in color. It was truly a tiny jewel in a wilderness.

The creek has silted up a great deal in recent years, and unless it is dredged will probably be completely filled in another one hundred years. Progress, in the form of eroded fields, woods, and soil, will ultimately destroy the creek and its surroundings. The people of that time will be the losers.

# The Ides of March

*The Ides of March have come.*

—Julius Caesar

The snow storms and lowering days of February have gone, and the Ides of March are on us. It is a good time to look around and see the forms of nature, including man, shuck off winter's lethargy and get ready for the resurrection of spring.

Dandelion leaves, toothed and green as a leprechaun's coat, peek out, take a look around, and almost overnight spread out to lengths of six to eight inches. This is the time to cut them for added spice in the tossed salad or cooked greens. Later on, gather the golden blossoms to make delicious dandelion wine. Many persons consider dandelions a nuisance, but the hardiness and propensity of the plant to vanquish all obstacles make it one of my favorites.

Trailing arbutus is one of the first plants to bloom in the spring. Look for it on a shaded hillside, and notice how the woody vine clings tightly to sandy, gravelly soil that nourishes the plant. The white and pink clusters of flowers are tiny and unassuming but carry a delightful fragrance. Coarse green leaves that tend to brown easily accentuate the beauty and fragility of the spring delight. Arbutus is the state flower of Massachusetts. Other names given the plant are shadflower, ground laurel, and winter pink. It is a rare flower in Virginia and should never be disturbed.

Creek and river shores, marinas, and scenic country landings, are alive with activity. Boats need to be repainted and their bottoms coppered to protect against the ravages of wind, water, and marine life. Weathered watermen, brown as old leather, check crab pots for needed repair and attach floats that act as Chinese lantern beacons to identify the owners and guide them along their routes.

Last spring I had the opportunity to tag along with Jack Smith of Naxera as he fished his pots. On several occasions I checked my watch to time him on his work. From the time his hands hit the float to bring a pot in, remove the crabs and old bait, put new bait in, and toss the pot overboard, only fifty seconds elapsed. He worked with dexterity that was truly amazing, learned through years of experience. I fished one pot in the ignominious time of six minutes.

Fishermen are preparing for another season. Today, nylon nets are used, and water people tell me they hold up better than the old style cotton nets that had to be tarred for protection. Even so, a large dogfish

14

or several voracious blues can tear up a net. Very little that is caught is wasted. Trash fish or mutilated fish go into crab pots for bait, while trout, bluefish, flounder, spot, and croakers are iced down and taken to market.

If March is warm and there are no blizzards, ospreys will be back from the south to look for nesting sites in and along the waterways of Tidewater. Favorite spots to build on are navigational buoys, and after several years' use a nest may reach eight feet in diameter. Ospreys are fiercely handsome, with a wingspread up to six feet. They are only now making a comeback after nearly being wiped out. The chain reaction of various insecticides that were ultimately ingested by the large birds when they caught and ate contaminated fish is generally thought to be responsible for the decimation of their numbers.

# The First Signs of Spring

*Oh, the gallant fisher's life!*
*It is the best of any;*
*'Tis full of pleasure, void of strife,*
*And 'tis beloved by many.*

—Isaac Walton, *The Compleat Angler*

The pickerelweed spikes that green up the marsh are among the first signs of spring, but are closely followed by the delicate white blossoms of the shadbush that nestles in the coves of the creek. Old-timers say that when the shadbush blooms it's time to get nets in the water to catch shad and herring that are making their yearly run.

Friends of mine start fishing then, and when they pull the two-hundred-foot nets, gleaming white sides of the trapped fish sparkle in the breezy sunshine. Spicy profanity, common to so many who work the water, sprinkles the conversation as the fish are pulled head first through the one and three-eighths-inch nylon mesh and deposited in galvanized tubs or buckets. The profanity normally is descriptive of an unusually large shad or herring, whose distended belly indicates the presence of the roe that is prized so highly. It describes also when the stiff-ridged spines of a dorsal fin are driven into fingers unprotected by gloves.

On the way back to the landing, a flock of Canadian geese flies over, headed north. They are unusually low, and again the air is salty. "Look at the *blankety, blankety, blank* geese. In hunting season, they never left Hog Island. Sure are pretty, aren't they?" The tones are strictly in admiration of the great birds.

At the house, the tubs and buckets are emptied onto a table, water is turned on, and skilled hands start to prepare the nearly two hundred fish for salting. Two men scale while a third cuts off heads and tails, and removes any roe that is found in the female fish. Woodrow is the cleaner, and he wields a razor-sharp twelve-inch knife with a speed and dexterity that is amazing to watch. It is not unusual for him to maintain a pace that keeps him up with the two scalers. I still wonder how he has kept all of his fingers for so many years.

# The Tree

*Poems are made by fools like me,*
*But only God can make a tree.*

—Joyce Kilmer

The immense tree looms alongside the nature trail like an aged sentinel. Tacked on its hoary side is a plain white sign with the number three on it. I look at the brochure and see the following words:

> This is a White Oak tree, approximately two hundred and twenty years old and since it is hollow in spots is probably a den tree for squirrels, raccoon, 'possum or possibly an owl. It is dead now but will probably stand for many more years unless it becomes a victim of a bad hurricane or tornado. The fruit of a white oak is called an acorn and is about an inch long. It is relished by many forms of wildlife including squirrels, deer, bluejays, wild turkeys and wood ducks. The Indians also used to bake the acorns and grind them for flour.

Is justice done by these words to this tall monarch that before succumbing fought for its life against fire, elements, and wildlife for so many years? The answer is a resounding no, and as I stand under it on this Sabbath day, I think of many, many other things that should be included.

I see it first as an acorn in the year 1750. A thrifty squirrel deposits the seed in rich alluvial soil but in the course of events fails to dig it up, and the tiny acorn sprouts and breaks through the matted forest floor. The damp soil bordering the creek waters the seedling, and for several years it grows rapidly. By 1775, the sapling is strong and tall, but it has been bent in an angle that will stay with it until the final crash to the ground. Was a trail-marking Indian or frontiersman responsible for this, or am I indulging in fancy?

During this same period of time, the tree also felt the winds of revolution that were born in Virginia, swept north to Massachusetts, and returned to blow themselves out on the beach at Yorktown, when Cornwallis surrendered and the freedom of our country was assured.

The time is now 1850, and rumblings are heard through the countryside. The rumble of talk will turn again to the rumble of war, and many men will pass through these woods. Is it not possible that the tree shaded or protected gallant soldiers of the Union forces who were striving to hold our country together? Could the tree with its distinctive marking have been a place of assignation for spies to meet or a social spot where opposing soldiers, minus officers, could exchange tobacco, sugar, or pleasantries?

By 1900, the tree is immense. From its sheltered spot, it can see lighters come up the creek to discharge cargo and to take on a load of logs. Giant cypresses and pine trees, neighbors of the big oak, fall to the broadaxes of brawny men in their fight to clear land for agriculture and to make a living by selling timber. Why was the oak spared?

Sometime prior to 1930, a lightning bolt flashed out of a summer night's storm, and with the booming of instant thunder hit the tree and buried itself at the base. In an instant, 180 years of growth were wiped out. In a week's time, all leaves had turned to an autumn brown, and by the middle thirties, top-most branches were starting to break off and fall to the ground below.

Today the tree stands, naked limbs stretched skyward, as if asking help from the Father who put it there. Its one-inch gray-green armor is cracking and falling off, and large hard-shelled beetles move slowly through the labyrinthine caves they have carved out. A cross vine moves up one side of the trunk, and a poisonous three-leafed ivy vine writhes its way up on the opposite side. Two giant roots are half exposed, and they are starting to decompose. Growing close by are cinnamon ferns and honeysuckle.

Despite its age, the giant will probably not fall for many years, and it will be at least a hundred years before it goes back to the dust that is its ultimate destination. In that period of time, it will continue to provide food and shelter to all forms of wildlife. Birds will perch in its branches and fly from there to seek food. It will serve to the end, and finally it will be, as in Ecclesiastes, XI, 3, "In the place where the tree falleth, there it shall be."

# Nesting Boxes

*Ne'er look for birds of this year
in the nests of the last.*

—Miguel de Cervantes

The month of March is an ideal time to build and erect nesting boxes for the numerous species of birds that inhabit Tidewater. Lots of natural home sites are destroyed each year by construction of new roads, apartments, and heavy industries, so to keep the bird life stable in a community, it is necessary for man to come up with substitute housing.

This practice has been followed for many years. Records show that the ancient peoples of Asia Minor used doves and pigeons for carrying messages, so certainly it was imperative to maintain adequate housing for these birds. Japanese temples provide shelves for swallows, and in India the Brahmans maintain feeding towers with nesting sites for the same types of birds.

In this country, early settlers found that the agricultural tribes of the eastern Indians hung gourds high in dead trees for purple martins to nest in. The colonists followed suit, and the practice is still carried on in rural sections of the country today.

Many birds have been recorded as using man-made homes on supporting devices. Some of the ordinary birds are wrens, blue-birds, chickadees, swallows, and woodpeckers of various kinds while the more glamorous include screech owls, barn owls, wood ducks, American goldeneye (duck), and hooded mergansers.

When a birdlover sees a wren build a home and raise a family in an old tin can, hat, or sleeve of a garment hanging in an outbuilding, it seems that almost any kind of structure would be attractive to the little fellows. This is close enough to the truth so that nearly anyone can build such a house and birds will use it.

It is much better, though, to take a little more time and follow a few simple rules. A well-built house should be durable, rainproof, cool, and have ready access for cleaning. This last rule is most important because most nests are infested with lice that should be destroyed.

The best material for construction of bird houses is wood. Metal should be avoided because it conducts heat that could suffocate small birds. Cypress, pine, and poplar are easy to work with, with cypress being the most durable. Slab wood (obtained from saw mills) with the bark still on makes a beautiful rustic home. If this type of wood is used, the cost is negligible.

If houses are made from finished wood, it is necessary to paint

19

them in tones of brown, gray, or dull green. Martin houses and any others placed in predominately sunny spots should be painted white to reflect heat.

Entrance holes should be placed near the top of the box, and the inner side of the lumber used should be left rough. This will assist the young to climb to the opening when ready to leave the nest.

Study the sketches shown with this article and try your hand at building a bird house. If you have no talent along these lines, get a woodworking friend to construct one for you. It takes only a couple of hours, and both of you can share the enjoyment of seeing some type of bird take up housekeeping chores in the home of their choice.

| Species | FLOOR OF CAVITY | DEPTH OF CAVITY | ENTRANCE ABOVE FLOOR | DIA OF ENTR. |
|---|---|---|---|---|
| BLUEBIRD | 5"x5" | 8" | 6" | 1½" |
| CHICKADEE | 4"x4" | 8"-10" | 6"-8" | 1 1/8" |
| TITMOUSE | 4"x4" | 8"-10" | 6"-8" | 1 1/4" |
| NUTHATCHES | 4"x4" | 8"-10" | 6"-8" | 1 1/4" |
| HOUSE WREN | 4"x4" | 6"-8" | 1"-6" | 7/8" |
| CAROLINA WREN | 4"x4" | 6"-8" | 1"-6" | 1 1/8" |
| CRESTED FLYCATCHER | 6"x6" | 8"-10" | 6"-8" | 2" |
| FLICKER | 7"x7" | 16"-18" | 14"-16" | 2½" |
| DOWNY WOODPECKER | 4"x4" | 8"-10" | 6"-8" | 1 1/4" |

# The Witching Hour

*It is a beauteous evening, calm and free,*
*The holy time is quiet as a nun*
*Breathless with adoration.*

—William Wordsworth

To some, the witching hour would be in the dark of the moon, with black clouds scudding across its face. Others would probably call the witching hour that time when a full moon is flushing out the corners of darkness and reflecting itself off every bright surface. To me, it is simply that period of time between six o'clock and dark on an evening in early spring.

In the language of today, I "goofed off" at such a time recently and was amazed at the many sights and sounds that I gathered. The seat I chose was an eight-foot section of a tulip poplar tree that had washed up several years ago and lodged on the bank. It had been a giant in its time but weather, worms, and woodpeckers had taken their toll. Nevertheless, it was comfortable and commanded an excellent view of the creek and surrounding woods.

The creek was in a placid mood, flat like a mirror except when a fish dimpled the surface, and the tide was moving in. Spring peepers had commenced a tuneful piping, and their sound engulfed the entire marsh. They are tiny frogs, difficult to locate since they cease calling when approached. Each time I look for one, I am reminded of a quote from John Boyle O'Reilly's *Rules of the Road.* "Be silent and safe— silence never betrays you."

Behind me a pair of chickadees were having a late meal at the feeder. They may have been newlyweds enjoying a night on the town or possibly the dignified couple that have rented my log house for the last three seasons. If this be true, they were probably celebrating the laying of a first egg in a nest made of leaves, moss, and straw, and lined with hair or fur.

A brown thrasher high in a lightning-blasted white oak tuned-up with a repertoire of songs that rivaled those of a white-winged mockingbird that was bursting its throat in a nearby holly tree. Starlings mewing from still another tree, and a nuthatch's nasal whine from back in the woods were punctuated by the regular crowing of my neurotic chanticleer. Most roosters crow early in the morning. Not this fellow. From his roost in a dense holly tree, he sounds off at any time— early morning daybreak, late in the afternoon, and middle of the night. During the day, he descends from his perch and condescends to mingle with his flock. Even then, at intervals, he flies up on a nearby post and

21

rehearses for his after dark calling.

As a rapidly sinking sun dipped the mares' tail clouds in gold and blue, two jets laid silver strands across the sky. They were so far away that no sound was heard to strike a discordant note in the sounds of nature, and the contrails spread rapidly to mingle with the clouds.

The first tree swallow of the season now appeared. It must have been an early scout, for it flew alone in search of early hatched insects. It was joined by a brown bat, whose erratic flight contrasted strongly with the smooth sailing of the swallow. Both would consume many insects in the course of their flight.

Grackles and red-winged blackbirds chattered their way home to a distant roost. A lone crow trailed along behind them, but as I watched, it dropped to a grove of pine trees. Here he was promised a peaceful night unless discovered by a horned owl.

Three great blue herons flew by, and one was in a shocking state of disrepair. Several primary wing feathers were missing, and I could only surmise that the great fisher had been the recipient of a shotgun blast or had possibly survived an attack by a stray raccoon or fox. Behind them, a fourth bird glided in with outstretched legs to find sanctuary on the marsh. Just before landing he sighted me, and like a plane being waved off a flight deck, took off with much wing flapping to try and catch up with the others.

A half moon was now directly overhead, and against the western sky a flock of large ducks was wedged in a flight that was pointed directly north. It was a fitting sight to close out the evening.

# The Lowly Dandelion

*Shed no tear—O shed no tear!*
*The flower will bloom another year.*

—John Keats

Springtime comes to Tidewater in the hurry-up rush of a behind-time train. Seemingly overnight, kelly-green buds of sassafras pop open to compete with colorful winged seeds of red maple. The white blossoms of shadbush along the creek appear, and the swollen buds of dogwood will soon give birth to the legendary white flowers that speckle the woods and forests of Virginia.

High bush huckleberry shrubs grow up to twelve feet and are loaded with minute bell-shaped flowers and small wedge-shaped leaves. The bark is reddish brown, and the July berries have few seeds and are tasty when taken from the bush or made into jelly, pies, or preserves.

Its small relative is called black huckleberry and is usually found on sunny, wooded slopes. In Canada, I have picked them along railroad tracks and in enough quantity to make several pies that were delicious.

Both types are classified as shrubs. They belong to the heath family and grow from the far north, to the Rocky Mountains, and throughout the extreme southern states.

May apples, or as I prefer to call them, umbrella plants burrow their way through the leaf mold and stand like partially folded umbrellas. A single white dot decorates the apex of the plants, and the delicate green leaves are heavily garnished with deep violet. Later on this shade will disappear into dark green leaves, whose color is accentuated by the single white flower that sometimes measures two inches. When the flower drops, a small green berry is formed that is reputed to be edible.

Legend says that this plant furnishes shelter for elves and leprechauns, and certainly it has the proper appearance for this function.

To most persons, dandelions are lawn weeds, but close examination will put them in a different perspective. With nearly two hundred brilliant yellow ray florets, they look like many miniature suns rising out of the earth. Toothed leaves are responsible for the name taken from the French *dent de lion* or lion's tooth.

Dandelions are extremely hardy plants. I have found blooms in every month of the year, so there is small wonder the plant is hard to eradicate. The heavy root drives deeply into the soil where it is safe from heat or drought, cold or extreme wetness. Rabbits, squirrels, moles, and insects pass it by because of the bitter taste.

For an added fillip to salad greens, cut dandelion leaves and throw them in a boiling pot of water with kale, turnip greens, or mustard. They are best in the early spring when young and tender. Reference sources state that roots can be dug, roasted until brown and crisp, and then ground up finely to make a type of coffee. I have eaten dandelion leaves but cannot vouch for the coffee.

When the yellow flower is gone, a globe of soft gray plumes appear. Children enjoy blowing these to the four winds and pretending that the number of puffs used to make the plumes disappear indicates the time of day. One puff, one o'clock, four puffs, four o'clock and so on. Each plume is like a tiny parachute ready to sail away to unknown destinations, and since they are seeds, they will germinate if landfall is made in a favorable spot.

I like this poem about the dandelion, from The Nature Library, "Wild Flowers" by Neltje Blanchan:

> Dear common flower that grow'st beside the way,
> Fringing the dusty road with harmless gold.
> Gold such as thine ne'er drew the Spanish prow
> Thru the primeval hush of Indian seas,
> Nor wrinkled the lean brow
> Of age, to rob the lover's heart of ease.
> 'Tis the spring's largess, which she scatters now
> to rich and poor alike, with lavish hand;
> Though most hearts never understand
> to take is at God's value, but pass by
> The offered wealth with unrewarded eye.

If I have made a case for the lowly dandelion, I'm glad. Too many times, the beauty of the commonplace is unseen by the beholder.

# Five-petaled Blossoms

*Small showers last long,*
*but sudden storms are short.*

—William Shakespeare

A spring storm is a beautiful and awesome thing. For several hours local radio stations mention its approach, and now the lowering clouds move into sight. The ones horizon-down are the color of blue marl and ominous in appearance. An occasional flash of jagged lightning splits the sky and accentuates the clouds' darkness.

In front of this air mass streamers of white clouds, like disembodied ghosts, scurry away before the wind that comes from the southeast. When massed together, a lively imagination can see a great fleet of ships with white sails bent before the wind.

Hastily the clouds merge and come in across the river to the marsh. Flocks of gulls, wings bright against the sullen clouds, sail ahead and take refuge in a small bay sheltered by high banks. They are joined by a gaunt blue heron whose heavy-going contrasts sharply with a pair of mallards that whip in with the wind, bank sharply against the turbulence, and land with bright yellow legs stretched out to absorb the shock of the water

Almost immediately, the rain storm hits. It is not a drizzly, drippy rain but one that commences hard and cuts off sharply. At its height, with a violent wind urging them onwards, the silvery slivers of raindrops drive in like needles. Surrounding trees with branches adorned in the spring finery of fresh budding leaves, bend before the blast to nearly sweep the ground.

All the songbirds have long since looked for shelter. I see a lone cardinal take refuge in an old attic and doubtless others are there also. There is no sound except the drive of wind and rain and I think of the many places a bird can hide. Hollow trees, stumps, bird houses, chimneys, awnings, eaves, empty buildings, and a multitude of other sanctuaries. A natural instinct drives them to such spots for self-preservation, and all other wildlife is endowed with the same knowledge.

The storm is short-lived. In an hour, the wind shifts to the northwest and wipes the clouds from the sky's face like a mother erases the tears of a child. The wind is still strong but has a cheerful voice that whispers through the trees.

On the river, whitecaps dance like silver dolls, and the gulls desert the protected waters to search for food. At this time of the year they fare well. Shad, herring, and perch are in the rivers and creeks, and those lost or injured by nets are easily taken by the graceful birds.

Next to leave the bay is the heron. It lumbers along and side-slips to land near a muskrat house in the marsh. At this spot it comes to attention with its long neck partially outstretched. I watch through binoculars and shortly a strike too swift for the eye to follow is made. A small fish about six inches long is impaled by the strong beak. The heron holds the fish momentarily and then swiftly gulps it down. The statuesque pose is again resumed while waiting for another prospective finny victim.

With the woods now bright and shiny as a silver dollar, I go looking for my favorite spring flower. They are hard to find but I remember patches seen last year. To locate the elusive trailing arbutus, it is necessary to travel hilly country and look on the shaded sides of steep banks. Here, intertwined with multi-green shaded mosses the little plant grows. The leaves, dark green, coarse, and heavy veined, hug the ground and the belled flowers nestle down among them. The five-petaled blossoms are generally white but sometimes range to a delicate pink. The flower is not over one-half inch in diameter, but the fragrance is unbelievable. Never should anyone pull a bouquet of these jewels. Instead, just pinch off a tiny blossom and be satisfied. Arbutus is rapidly becoming extinct, and extreme care must be taken to preserve the few remaining patches. It all of them go, one of the most enchanting sights of early spring will be denied us, and surely we will be the poorer for it.

# May

*We're never single-minded, unperplexed,*
*like migratory birds.*

—Rainey Maria Rilke

Each season has a special month that exemplifies all that is good, and May is certainly representative of the best of spring. The chilly days of April have been swept away by the Shawondasee* wind and millions of birds are winging their way to hereditary nesting sites.

Longfellow was obviously very aware of springtime when he wrote the following lines from "The Song of Hiawatha:"

Shawondasee, fat and lazy,
Had his dwelling far to southward,
In the drowsy, dreamy sunshine,
In the ne'er ending summer.
He it was who sent the wood-birds,
Sent the Opechee, the robin,
Sent the bluebird, the Owaissa,
Sent the Shawshaw, sent the swallow,
Sent the wild goose, Wawa, northward.

May is when the trees, resplendent in new outfits of shimmering greens, are alive with birds; feeding, singing, choosing mates, or simply resting for the next leg of their journey. Many stop and nest in Virginia, but others will continue to northern points of Alaska, the Yukon, and the Arctic Circle. The Canadian provinces and Alaska are considered the greatest nursery for waterfowl in North America.

Many birds, including the tiny hummingbird, make a nonstop flight across the Gulf of Mexico. This five-hundred mile trip is made in a single night so that food is available before departure and can be obtained immediately upon arrival. Another long flyer is the golden plover. Weather permitting, it is known to fly from Nova Scotia to South America nonstop, a distance of twenty-four hundred miles in not quite forty-eight hours. Fuel used is estimated to be less than two ounces in the form of body fat. If scientists could unlock this secret, our energy crisis would be nonexistent.

The arctic tern is the world's champion in migration habits. It nests as far north in the Arctic Circle as land can be found, with the first recorded nest being located only a short distance from the North Pole. It

---

*Chippewa Indian word meaning southwind.

usually leaves the nest around the last of August and shortly thereafter moves to the edge of the Antarctic continent, a distance of eleven thousand miles.

Some birds migrate by day, notably swallows, swifts, ducks, geese, and hawks. The swallows and swifts feed in a zigzag flight towards their final destination. Most birds, however, seem to prefer night travel so they can feed and rest during the day. Many night flyers lose their lives by flying into lighthouse and airport beacons that dot the coastal east. This is especially true in foul weather.

It is interesting to ponder how birds find their way in migrations and an answer to suit everyone has not been discovered. Certainly it is true that the same birds return year after year to the same nesting places. In my personal experience, chickadees and phoebes have returned to identically the same sites for four years. Do they have an inner radar that guides them? Do they follow certain physical roadways such as rivers, mountain chains, or coast lines? Possibly. The answer probably is a little of both, with a third major factor being available food supply.

Some years ago, I had an interesting experience with a migratory bird. It was a small bird that was feeding in the top of a sweet gum tree. I followed and observed it closely through binoculars for nearly thirty minutes so positive identification could be made. I finally determined it was a blue-winged warbler and added it to my life list of birds. The next day at work I called a friend who is an outstanding ornithologist and told him what I had seen. As an old buddy of mine, I could tell by the way he talked that he was just a bit dubious of my identification. The next day, however, he telephoned me and said that his wife, who is also an outstanding birder and lecturer, had met him at the door when he came in and told him that she had sighted and identified several of the same species in their back yard. I was delighted to be vindicated in my identification.

# A Phenomenon of Nature

*Only the gamefish swims upstream.*

—John T. Moore

Recently, I had an occasion to visit a reservoir spillway where I was treated to a phenomenon of nature. The day was one when Lady Springtime showed her best behavior. A soft sky, shaded like a bluebird's wing, was highlighted by great masses of cottony cumulus clouds drifting through the heavens. Clusters of white fringe trees that sometimes trailed branches in the water accentuated the dark green of oak, maple, and hickory along the shoreline. Turtles basked on logs in the warm sunshine and a water snake wove a path across the lake.

The spillway was of ancient vintage. Eleven courses of concrete, each three feet wide, were stepped from the holding pond to the down-water stream. Tucked in the corners away from the wash of water, scum had collected over the years. It added an emerald green touch to the yellows and browns built up by mineral deposits that adhered to the surface of the concrete. Unless memory from long ago biology days plays me false, the greenery was spirogyra, more often called "pond scum." It can be seen by the naked eye to consist of a great number of interlacing threads, and when studied under a microscope shows the threads to be a series of cells joined end to end. It is slippery and slimy to the touch.

The water rolled swiftly from the big pond down over the ladder-like arrangement of concrete steps and emptied into a deep body of water at the base. This was the spot where the miracle of springtime was taking place, and I could hardly believe my eyes as I watched the sprawling mass below.

Literally hundreds of fish, mostly carp, were there. They rolled, splashed, and frolicked in the violet-shaded deep water like a playground of school children. Some were small, some medium, and many were immense brutes. The large ones were nearly three feet long and weighed probably fifteen pounds. Dark backs, topped by heavy dorsal fins, rose high out of the water as they moved about.

Many of them had a single urge, and they took turns trying to fulfill it. At that moment, swimming, climbing, or leaping up those concrete steps seemed to be the most important thing in the world to them. Since the water on the steps was running swiftly, and not over two inches deep, the efforts of the big fish were ludicrous.

The big ones took turns. They approached the spillway with the dogged determination of a trained athlete and heaved their massive bodies strongly upwards. Some made it, landing on their sides or bellies. Huge black scales shimmered the sunlight, and yellowish stomachs flashed as the successful leapers strove to maintain position for an assault on the second step. It was never to be. Like an inexorable force, the rippling water each time repulsed the jumpers but did not deter others that awaited their turns.

Occasionally a smaller fish would attempt the jump. They wriggled their tails in a valiant effort to keep going, but lack of stamina and rushing water inevitably swept them back to the deep water.

While this was going on, formations of four, eight, ten, and sometimes larger groups swam in the precise manner of a flotilla of ships, moving under the direction of one master. They banked, turned, wheeled, and maneuvered with the balletic grace of a flock of blackbirds or starlings and, like the birds, apparently with no communication except instinct. It was a sight I will long remember. When I returned home, a reference book gave me a few interesting facts about carp.

They were introduced into this country in 1872 and have since spread to most lakes, streams, ponds, and rivers. They came originally from the Far East, traveling the same road with western civilization.

Carp are noted for longevity. Natural history works of the eighteenth and nineteenth centuries attribute to them a life span of up to 150 years. This is now considered a gross exaggeration, and in the wild state, carp are not believed to live much over fifteen years.

Izaak Walton called carp a "water fox," possibly because of its fondness for feeding almost exclusively on vegetable matter that it roots from the bottom, thus making it a difficult fish to catch. Grain, like corn or wheat is considered the only bait that will catch carp successfully. When caught, the fish puts up a good fight.

The large fellow is popular as a game fish in Europe, where it graces many tables. Probably at some future date it will come into its own in this country and prove to be a desirable fish to take home.

# The Winged Enemy

*To own a bit of ground, to scratch it
with a hoe, to plant seeds, and watch
the renewal of life—this is the
commonest delight of the race, the most
satisfactory thing a man can do.*

—Charles Dudley Warner

Today I have been attacked and routed from my garden. My back, arms, and legs are blood-splotched, and my bald head feels like headquarters for a convention of chiggers, or in the terminology of the deep South, redbugs. I am irritated, mad, disgusted, exasperated, annoyed, and slightly profane. My wife decides that very definitely the grandchildren must be returned to their home immediately. I, like Achilles, am left to sulk in my tent alone.

Let me tell you how it all came about. The day is late in the spring, and vanilla ice cream clouds are floating through a bluebird sky. The temperature is about eighty-five degrees, and many birds are tuning up for the serenade that will follow later in the afternoon. A turkey buzzard wheels around the sky looking for carrion far below while swallows and swifts cut swiftly through the sky in pursuit of insects. It is a beautiful day.

This is when I make the decision to pick garden peas. The vines that have been cherished, babied, and nursed since early in February are heavily laden with chartreuse-colored pods that hang in clusters and are so swollen with tasty peas that they look ready to pop.

My gardening habit consists of shorts, socks, and L. L. Bean hunting boots. I am a sun worshipper and firmly believe that the rays of Old Sol are good for colds, arthritis, headaches, rheumatism, and chilblains, so I am minus a shirt of any kind. In my right hand I carry a two-and-a-half gallon plastic pail in which to put the peas.

On the first row, I place the pail about two feet away, assume the proper bend-from-the-waist position, and start to pick. Almost immediately I am nailed on the upper thigh by what had to be a flying hypo needle. Before I can straighten up, its wingman delivers a Sidewinder missile that penetrates my left forearm. Trying to be stoic under attack, I continue to pick but very shortly I start my counterattack when the squadrons commence to use my head for landing and take-off practice.

It is partially successful, and I down two of the enemy. To my surprise, they are not warplanes or missiles but tiny flies, and I remember that during the months of May and June they are torment on earth. The correct name of the pests is deer flies, although many call them mayflies. Still others, including farmers, fishermen, and outdoorsmen, have more descriptive names that cannot be repeated here.

31

The insects are less than one-half inch long, with two silver wings that have a black blotch of color near the center. The females are equipped with a long beak that oftentimes draws blood.

For one hour I pick peas and fight flies. They come in swarms or singly. Some hum loudly, and some fly in on noiseless wings like gliders. All have one objective: get as much of my blood as possible, then come back and strike again. I pull a large bandanna handkerchief from my pocket and tie it on my head so I resemble the Sheik of Araby. This helps for a while until they discover that my ears are orifices to be penetrated and explored. By this time I am a maze of welts and bites that sting, itch, and tingle. The warm day has made me perspire and slap the enemy with dirty hands, making me grimy. Three rows of peas remain to be picked, but that chore must be performed in early morning hours when gossamer wings are still damp with dew and the winged enemy is grounded.

Not long ago I read that scientists are formulating plans to cut down on the fly population by sterilization methods. I am not in favor of seeing any species of wildlife wiped out completely because of the possible disastrous chain reaction in nature. But certainly I would not be averse to the elimination of about five million deer flies from my cotton-picking pea patch.

# Fishing

*We catched fish and talked.*

—Mark Twain

I recently went on an outing altogether different from other pastimes. It occurred in the declining days of spring and happened quite by accident. If you have not been on a herring-dipping expedition, and I had not, perhaps a few words will get you the notion to give it a try during the month of May.

Four of us had been riding on a Sunday afternoon, enjoying the newly-awakened spring flowers, singing birds, and the trees in their pastel-colored leaves. When we saw a number of cars stopped at a small bridge under which a light-catching stream was rambling, our immediate reaction was to stop. Stationed along the stream that glittered its way through a tangled swamp were lots of people holding long-handled nets in the water. Ever so often, the nets were hoisted out, filled with wriggling fish. They were herring, on their annual spawning journey that takes them up every small rivulet or watercourse off a tidal river. They were rakish-shaped fish with silvery sides that caught every ray of the afternoon sun as the lucky fishermen stowed them in buckets and ice chests.

Questioning of the fishermen gave us all the needed information. A county permit costing two dollars, a net, and a fish container were all the items required to get into the serious business of herring-dipping.

Late in the evening of the following Wednesday found us outfitted and located at a likely spot on the stream. The tide was making out, and few fish had been caught by other fishermen. Night falls rapidly in a deep swamp, and myriad sounds fill the air. A bullfrog chunked from a nearby log, insects of all kinds hummed a lullaby, and in the distance a lone owl hooted.

As we sat in anticipation, a large-barred owl drifted in on silent wings and landed in a tree just above us. Apparently we were not seen because it sent out a hunting call immediately. Shortly it fluttered down to a lower branch where cruel brown eyes shot a vicious glance at what appeared to be a knotty branch. A catapult launch by the predator revealed the knot to be a panicky female wood duck, which had been reached, but not caught by the curved talons of the owl. She must have flinched at the blood chilling call and this small movement had been picked up by the alert eyes. The duck broke free and fled in headlong flight, hotly pursued by the owl. Just a short distance away, the small flyer splashed down to the safety of a log jam, and the killer flew off to resume its hunt for other prey.

This particular species of owl does not make a habit of killing such game, and I feel this was an isolated case. Most of the time the victims consist of mice, frogs, lizards, and larger insects. It was the first time any of us had seen an attempted kill by an owl, and needless to say, we were glad to see the duck escape.

In two hours time the tide swung around and fish started to come in with it. Each net was kept submerged and as the herring swam in, the net would jiggle much like the feeling a fisherman gets when holding a hand line. It was exciting and fun.

Lifting of the net at such times resulted in catching anywhere from one to a dozen fish. In the opaque light of a now visible moon, the fish shimmered sparks of white lightning. They twisted, turned, and thrashed in the confines of the net and sent showers of water spray to saturate an already muddy bank. Some hung in the mesh and could not be shaken out. This entailed working them head first through the net, often at the expense of pricked hands from sharp dorsel fins.

At two o'clock we counted a total of 109 herring. They were still running in the now high-level stream, but visions of cleaning our catch helped us decide to quit. Navigating the five hundred yards of slippery, gooey trail back to the car by flashlight was accomplished without mishap.

By the time home was reached, clouds had obscured the moon and rain was falling. Some lightning was in the storm, so paper was spread on the kitchen floor and the work of preparing the fish for eating was begun. When midnight arrived all the herring had been scaled, cleaned, washed, and salted down in plastic pails. Many with distended bellies had yielded enough roe to fill two pint containers. The heads, scales, and internal organs were saved to be placed under tomato plants for fertilizer.

Herring are small. Herring are bony. Herring are also lots of fun to catch. The true worth of them, though, will be found when they are fried crispy brown and eaten on a cold winter morning. Cooked this way, all bones except the back bone may be eaten. The one hundred fish would carry us through the winter.

After cleaning, large tins are brought out, and the fish are heavily salted and layered in the tins. Prepared in this way, they keep until winter, and as Woodrow says—and I agree with him—"There's not much better on a cold morning than hot biscuits loaded with butter, coffee, and a herring fried so crisp that you can eat all of him, including bones."

Herring and shad are very important food fish. They belong to the same family as menhaden, sardines, and alewives. Herring is one of the more numerous of all the backboned animals, and it is not unusual for millions of them to swim close together in schools.

Most of their time is spent in deep ocean water, but in the spring they migrate up the smaller coastal waters where eggs are deposited.

One female will deposit up to sixty thousand eggs, but due to natural enemies only a few develop into adult fish.

About ten trillion herring are caught each year, with Norway leading the way in the number taken. The fish are frozen and used for bait; oil used for lubricating machinery is extracted from them after which the remains are ground up for fertilizer. Large quantities are pickled and smoked for eating purposes, with kippered herring probably being the best known.

# The Six-mile Route

*A proper man, as one shall see in a summer's day.*

—William Shakespeare

Bicycling is fun and is an excellent form of exercise. Much more ground can be covered than by walking, and yet it is no trouble to stop and visit with persons along your route or to see the wildlife in the woods and fields that are passed. There is also one other advantage to cycling: one seldom comes home empty-handed.

I ride a three-speed bike and usually follow the same six-mile route. My special preference is late afternoon riding when traffic is sparse and I can concentrate on the countryside. Blue jeans, short-sleeved shirt, tennis shoes, and baseball cap are the most comfortable clothes to wear for spring, summer, and fall riding. In the winter, add thermal underwear plus a hooded sweatshirt. Thus garmented, a bicyclist can generally ride in any kind of fair weather. I also carry binoculars, notebook with pencil, and a large feed bag.

My most recent outing was in late spring. There was no wind, the temperature was in the low eighties, and the sun had started a slow descent in the western sky. High cumulus clouds were changing from daytime white uniforms into colorful evening clothes of saffron, orange and violet. They hung in the sky in suspended animation as if awaiting a push to start them to a party.

As I rode, mockingbirds were much in evidence, as were common grackles. In the still air the long-tailed grackles did not experience the difficulty in flying that they do in heavy winds. As I wheeled along, a yellowish-colored bird started from a thicket to fly into other nearby bushes. I did not recognize it, so I braked to a stop and grabbed my binoculars. When I lifted the glasses, the bird moved out to a limb and posed in a picture-taking stance. It was a yellow-breasted chat, oftentimes heard but seldom seen without going into the bushes after it.

Further on, a four-foot king snake slithered across the road but I was not fast enough to catch it. This is the function of the feed bag mentioned earlier. Any black or king snake or unusual kind of turtle or terrapin that I can capture is transported via the sack back to my woods where they can be turned loose and live in comparative safety.

Rabbits were seen, mostly adults, but some young were visible in the recently cut weeds. Two were ragged looking and I assumed them to be female that had pulled fur from their breasts to use in lining a nest for newborn young.

36

My next stop was to say hello to friends known for many years. We chatted for a while, after which they showed me their strawberry patch, where crimson berries the size of plums were nestled in beds of vivid green leaves. Not a blade of grass was evident, and I asked the secret. Mulch in the form of chips and wood shaving, I was told, plus a little hand pulling.

They next took me to three cherry trees that were loaded with clustered globules of red fruit. The cherries glistened with brightness and reminded me of a Christmas tree I once decorated with red ornaments only. Grackles, mockingbirds, robins, and cardinals were active on the outside highest branches, and the ground was littered with discarded pits.

When asked if I would like to pick a box of the fruit, my answer was naturally affirmative. Visions of hot cherry pie topped with vanilla ice cream floated in front of my eyes, and the box was soon filled. When the distaff member of the family urged me to pick more, I was reluctant until she said that she had picked, canned, and frozen cherries until she was sick of them. When the three of us stopped picking, it was nearly dark, and I had four quarts of cherries and one quart of strawberries.

My friend is the kind of fellow who can improvise, and he shortly came up with a carrying case that I could transport the goodies in while riding the bike. I thanked them and left as the sun was just dropping behind the horizon in a colorful sky show. In ten minutes I was home. As I said earlier, a bike rider seldom comes home empty handed, be it snakes, turtles, or cherries.

# Spring Recipes

## Plumbiere

Make a rich custard (sixteen eggs to three quarts of new milk); when cool, sweeten to taste and allow one tablespoonful of sugar to an egg; add one and one-fourth pounds seeded raisins, and one-fourth pounds almonds blanched, and cut up; add three-fourths pound citron cut thin and into pieces half-inch long; season custard with vanilla; whip half gallon cream and mix. Freeze it very hard.

## White Taffy Candy

Mix six pounds sugar, half pound of butter, one teaspoonful cream of tartar; boil until it is brittle, pour out, and pull white; season with vanilla.

## Fish Cakes for Breakfast

Take any fish you may have left and pick clean from bones and skin. Chop twice the amount of boiled potatoes. Add one egg. Make into small cakes and fry to light brown in hot drippings. Take care not to burn.

## Strawberry Souffle

Take one quart of fresh strawberries, mash them through a colander, add one cup sugar and the beaten whites of five eggs; place in buttered dish; sprinkle sugar over top; bake slowly half hour. Serve with cream sauce.

## Old Virginia Ginger Bread

Mix two quarts of flour, one pint of molasses, two pounds coffee sugar, three-fourths pound butter, and lard mixed, five eggs, one teacup of sweet milk, four teaspoonfuls baking powder; make the batter stiff; butter the tin pan and bake the cake slowly.

## Pineapple Sherbet

Use two cans of pineapple; chop very fine; add to this three pints of boiling water and the juice of three lemons; sweeten to taste and freeze hard.

## Duchesse Potatoes

Bake six potatoes until done, with their skins on; then take out the inside carefully, and mash them through a strainer. Beat up two eggs, yellow and white separately; one cup milk or cream, large lump of butter; put the milk in a saucepan; stir in the potatoes, then the eggs, yellow first, then whites; butter and salt; when it gets hot, put back in the skins, and bake light brown.

## Fried Oysters

Use fine oysters, drained well through a sieve; have egg well beaten; salt and pepper the oysters and dip into the egg; roll in cracker dust, and fry light brown in boiling lard.

## Jubal Early Punch

Take one and one-half gallons of lemonade; three pints Cognac brandy; one pint of rum; three quarts champagne; dissolve the sugar in the lemonade, allowing one and one-half pounds. Serve with lots of ice.

## Oyster Soup

Strain half-gallon of oysters; rinse them through two waters. Take the liquor after straining it, season with two small onions chopped fine, three stalks of celery (chopped), five or six blades of mace, pepper (red and black), and salt. Let this come to a boil; then add the oysters; let them cook until the gills curl a little, then put in one quart of cream or milk. Thicken the soup with two tablespoonfuls flour, rubbed into four ounces of butter. Serve hot with crackers.

## Damson Pie

Mix five eggs (beaten separately), one teacupful of damsons, one teacupful of sugar; flavor with vanilla; bake in thin crust. This quantity makes three pies. Fresh fruit should be added.

# Dandelion Wine

Pick the blossoms early in the morning. Gather enough to make two quarts of petals after stem and green collar are cut off. Rinse blossoms in cool water before preparing petals, then place in a large pan and cover with four quarts of water. Boil the flowers for twenty minutes. Pour the hot liquid and petals over two oranges and two lemons cut into small pieces. Allow mixture to cool to lukewarm. Add one yeast cake and let stand for forty-eight hours. Strain mixture through cheesecloth, squeezing to remove all juice. Add three and a half pounds of sugar to juices and stir well to dissolve. Pour into jug with a lid but don't screw the cap down tightly. The wine should stand for about six weeks or until still. Strain and bottle. Keep for six months at least before drinking. This makes about five four-fifth wine bottles.

# SUMMER

# On This Day

*And what is so rare as a day in June?*
*Then, if ever, come perfect days;*
*Then Heaven tries the earth if it be in tune,*
*And over it softly her warm ear lays.*

—James Russell Lowell

It is real hard to beat a day in June. The whole world seems to be alive and on the move. Especially nice are the occasions when no rain has fallen for a couple of weeks. Suddenly the wind swings around to the northeast. Just as though it were released by a trigger, the instant air conditioning of nature cools the atmosphere. Mare's tail clouds appear on the horizon and send long streamers ahead of an approaching front. The sky of robin's egg blue is erased by clouds that gradually melt into one big mass. That brings welcomed rain.

With built-in weather barometers, birds, animals, and insects sense falling weather and react accordingly. Squirrels and rabbits scamper around the clearing to feed. In broad daylight a raccoon meanders along the creek and leaves baby foot tracks in the mud. A muddle of his tracks and a small pile of empty mussel shells will remain to show that "Kilroy was here" until erased by a rising tide.

The surface of the creek is alive with insects. They skim and dart along the top of the water in leisurely fashion until I approach to get a better look. Then, like jet pilots, they turn on extra fuel and scoot to the sanctuary of deep water. They range in size from nearly microscopic to one-fourth an inch. Looking down on them I can see they have four legs, two antennae, and black cigar-shaped bodies striped with yellow. I am no entomologist and from my small insect book can only determine they are a type of water treader. The tiny insects are not being harassed by fish, but rough-winged and barn swallows are having a field day. Dozens of the birds are swinging and circling over the creek and surrounding marsh. The rough-winged swallows with brown bodies and cream-colored waistcoats are predominant and swoop in low-level attacks that nearly touch the water. As I watch, a large bass strikes with a tremendous splash at one of the low flyers. The swiftly moving bird is gone, though, and only ripples hide the spot where the predatory fish lurks.

Occasionally, the birds tire and fly to the top of a partially dead cypress that leans in precarious fashion over the creek. While resting, they preen themselves and twitter in high-pitched squeaks. When forked-tailed barn swallows trimly dressed in steel-blue top-coats and buff cummerbunds land near the soberly dressed rough-winged birds, the contrast is great. Never were two birds of the same species so unlike.

As they sit and chat, a quarrelsome kingbird flies up and, just like an overseer, quickly disperses them to get about the job of catching insects. It is no novice at such work and in the space of a few minutes he catapults off the branch several times to snatch larger insects that fly by. Each time a snapping of the beak indicates another bug caught, and since this act occurs hundreds of times a day, one realizes the immense value of these birds to man.

The kingbird is a take-charge guy and tolerates no approach of crows, hawks, grackles, or other large birds. Many times I have seen this member of the flycatcher family nearly attach itself to the back of other birds to drive them off. When it flies from one perch to another, the white band running across the fan-like tail makes identification easy. Not so easily seen is the patch of red feathers that adorns the crown.

A brown ball of feathers streaks across the marsh in headlong fashion. Even at a distance the speedy flight indicates bobwhite. Shortly its ringing call is heard to summon the female that probably flew in a different direction when flushed.

On this day, the cypress with dead branches that stretch like arms across the creek is a haven and lookout spot for several kinds of birds. As I write, others birds besides the three mentioned have perched there. A bluejay stops by for a while as do two red-winged blackbirds. Another flycatcher takes over in the form of a phoebe, and a cardinal and mourning dove rest there for a few minutes. A large grackle, feathers iridescent in the bright sunshine and long tail streaming out behind stops, looks around and then moves on. A pair of prothonotary warblers chase each other around the tree trunk, while a downy woodpecker grubs for worms in one of the dead limbs.

The tree is listing badly and has for many years. It has seen much action on the creek, both human and wildlife. Its roots go deeply into the black mud, and around the base can be seen the remains of a slab runway where logs were loaded on ships many years ago. Here an occasional water snake slithers from the water to sun itself, and turtles find shelter under the bank. Someday a big storm will topple the tree, but the water is not deep enough to submerge it completely, and it will continue to serve for many years as a perch, lookout, or sunning spot.

# The Wonders of Nature

*Nature does nothing uselessly.*

—Aristotle

The late evening is still. Not a breath of wind moves the branches of trees bordering the creek, and there is no ripple on the water. A golden sun, low in the western sky, is attempting vainly to hold back the purplish shadows that move across the marsh. It seems a perfect time to get some exercise with the fly rod, so I assemble it rapidly, grab the tackle box, and push my small aluminum boat into the water. The tide is coming in so I paddle about half a mile against the tide and then start the slow drift back.

I have cast only a few minutes with no results when I realize I am about to witness one of the miracles of nature. Mayflies are hatching, and when this occurs the fisherman might as well lay his fishing rod down unless he possesses an artificial mayfly or a reasonable facsimile. I do not have the proper lures so I sit and enjoy watching the hatch as much as if I am catching several fish.

Deer flies are often called mayflies, but the ones I watch now are true mayflies, prettier and more graceful, and not the biting pests that will come in early May and June. Since there are thousands now in the air, I am glad that these winged creatures lack a mouth or stinger with which to attack.

As I drift slowly along, a dimple appears on the surface of the water, and in less than ten seconds a full-grown mayfly takes to the air. Its empty nymph case then floats on into oblivion, its task completed. The fish are feeding voraciously on the nymphs before the insects actually hatch, and if the large-winged bug is slow getting off the water, it is quickly eaten. Once hatched and airborne, it instantly becomes the prey of red-winged blackbirds, kingbirds, fly-catchers, and bats.

The most unusual birds I see feasting on the flies—and probably the ones I would least have suspected—are two yellow-billed cuckoos. It is comical to watch them swoop from the high-limbed cypresses, sweep across the marsh, and grab the flying tidbits. The long tails, broadly washed with white, semaphore their identities, and they fly with a swiftness and openness totally unlike their usual secretiveness.

Mayflies are buff colored and may be as long as two and a half inches. They are delicate creatures, having four wings nearly transparent, and two or three long appendages projecting from the end of the abdomen.

The instant they hatch and are airborne, they seek protection of low-hanging bushes and branches. When the hatch is at its height, thousands of them are in the air at once, moving with the balletic grace of tiny fairies.

There are one hundred known species in this country and over five hundred in the world. Many of these insects have been found in a fossil state.

Adult mayflies live only a day or two, and during this time sexual union takes place. Each female will deposit five hundred to one thousand eggs in fresh or brackish water, after which she dies. When the eggs hatch they sometimes take up to three years to complete a life cycle.

The nymphs live under water, burrowing in mud during winter and under stones and logs during the warm months. Unlike the adults, nymphs have the capability to chew and feed primarily on plant material.

They are able to stay underwater because of gills alongside the abdomen.

Mayflies are valuable as food for fish and do not seem to represent any danger to mankind despite their numbers. In the world of nature, they have a place, and it is a pity that more people do not have an opportunity to watch a mayfly hatch on a warm summer evening.

# A Growing Rain

*Ye marshes, how candid and simple*
*and nothing-withholding and free.*
*Ye publish yourselves to the sky*
*and offer yourselves to the sea.*

—Sidney Lanier

Today we have a growing rain, and it is good. Soft raindrops dimple the creek and enter the earth, softly and without fanfare of lightning or trumpeting of thunder. Silver drops wash the dust from corn, potatoes, beets, and other garden goodies as well as the plants and flowers of the woods.

Just before dusk I don hip boots and foul weather gear and walk out into the marsh. Pickerel weed, with arrow-shaped leaves, are two to three feet high and growing so closely together that I cannot see the ground when I place my feet down. In the last several years, most of the cattail growth has disappeared from the marsh, and its place has been usurped by these plants. A carp beats a hasty retreat from a nearby gut, where it has been feeding, and the wake indicates a fish nearly three feet long. The water is so shallow that mud is stirred up by a fish that is unpopular in this country yet sought after in European countries.

Farther along in the marsh I find what I suspected would be in bloom. It is the larger blue flag and is similar to the common garden variety of blue iris. The violet sepals are heavily veined and splashed on the lower portion with a dash of buttercup yellow. I check my reference book for terminology and find that the three stamens are overlapped by three petals that are called "styles." This is new to me and makes me glad I brought the book. Otherwise, the iris would have had nine petals and I would have given out false information. Anyway, it is a beautiful flower and well worth a search.

As I walk on, a male wood duck flies up with a plaintive squeal. I am sure the female is in a nearby hollow sitting on eggs, so I move quietly away. Cypress knees are plentiful in the swamp and in the heavy growth, and several times I nearly trip over one.

Along the edge of the swamp, great masses of cinnamon ferns are growing. This is the only fern that I can identify with certainty, and I remind myself that I need a book on ferns so I can learn more varieties. The cinnamon is unmistakable with its toast-colored poniard that projects upward from fronds twelve inches wide and as much as forty inches long.

The rank growth of poison ivy is apparent throughout the marsh, and it is particularly toxic at this time of year. The wet new leaves are June apple green, and the plant is in full bloom, with clusters of three

48

leaves. In bush form it is known as poison oak, but by either name it should be avoided. In fall, the leaves turn crimson or yellow, and masses of poisonous drupes like mistletoe berries appear.

During the course of my walk I come across a beautiful specimen of tree. It is a fringe tree, and the first one I saw reminded me of a silvery mist, floating on branches. The ones I find today are really a cluster of several trees. They are about thirty feet high and twenty feet across. At a distance the blossom appears as a froth of fancy white lace, set off by clusters of green leaves shyly peeping out.

The fringe tree is used as an ornamental tree or shrub and can be grown in rich, moist soil. The white flowers are composed of four narrow petals one and a half inches long that are separated nearly down to the base. Forsythia, jasmine, and olive are other members of the family.

As I head back to the house I see columbine, wild ginger, and blackberry vines loaded with clusters of snowball white blooms that will develop into blackberry cobblers if the weather is seasonable. I also notice that the lovely white flowers of the may apple or umbrella plant have dropped and been replaced by dwarf, watermelon-shaped fruit. This forcibly brings to mind a quotation from Immanuel Kant: "Everything in nature acts in conformity with law." I return to the house satisfied.

# This Forgotten Pasture

*Remove not the landmark
on the boundary of the fields.*

—Amenemope

A deserted and grown-up field is a panorama of sights, noises, half-heard bird songs, and constant buzzing and chirping of hordes of insects. It is an area where nature rules with a bountiful and luxuriant growth of its creatures. Sad to say, there are not too many fields of this type for the average person to enjoy.

Recently I sat in a grove of pecan, black walnut, and hickory trees that still remained in a remote field. There were seven in the grove, and they were growing in a semicircle like sentinel outposts guarding the cornfield. Scattered throughout the pasture were other nut trees, and in the distance an ancient red cedar, gnarled and weather-whipped, spread dark green branches in traditional Christmas tree fashion.

The hickory tree was noteworthy. It was a good thirty inches in diameter and soared skyward about eighty feet. I read in Annie Dillard's book, *Pilgrim at Tinker Creek*, that a large elm tree has in the neighborhood of six million leaves. This was an unusually large hickory, and nearly all its limbs show a perpetual bow from the weight of multitudinous leaves. I will accept Annie Dillard's information as fact and not try to count the leaves on this giant.

The leaves are nine inches long and five inches wide through the middle. The stem is heavy at the base and as thick as a wooden matchstick where it is affixed to the branch. The leaf is heavily laced with ocher-colored veins that weave sinuous lines from the main stem to the edge of the leaf.

I examine several dozen leaves and do not find a perfect one. Insects that were dormant or still in a comatose state when the leaves first appeared have taken a heavy toll. Holes, jagged edges, and scale, brown as a mink's coat, have ruined the perfection of the spring leaves. The corpse of a honeybee, wings stripped away, is fastened to the underside of one leaf with a gluey substance that defies removal.

The tree is loaded with nuts the size of a small tangerine or large plum. The hull is thick and green as an unripe apple. When I open the four segments with a pocketknife, almost instantly the inner portions turn from milk white to the color of a crook-neck squash that has been warmed by the suns of summer. The actual nut has not attained full growth, and the shell has a tint of pink.

Throughout the field are heavy growths of honeysuckle and three-fingered poison ivy. Crab grass is evident, with seed spikes four feet tall. At this time of year the common horse nettle is blooming profusely. Scorned and despised in the garden where tiny thorns prick unwary weeding hands, the five-petaled, white-to-blue tinged, star-shaped flowers are beautiful in this forgotten pasture. I pick one to study the fragility of the delicate petals and immediately see a tiny black six-legged creature clambering over and around the inverted cones representing the anthers. A knobbed stem wearing a Robin Hood cap leans over the brilliantly hued anthers like a Lilliputian preacher exhorting his flock to mend their ways. Later this stem will produce an orange berry.

It had been raining in light showers, but the dense overhead foliage protects me. At first, flies and insects are numerous, but moving to shelter under a black walnut tree eliminates this problem. It is a fact that few bugs of any kind can tolerate staying under this species of tree for any length of time. Apparently the tree exudes an odor that repels them. Cows and horses gravitate to this kind of shelter in summer when insect pests are so worrisome.

The field is alive with birds in the early morning. Bobwhites talk to each other by long distance, and a Carolina wren almost bursts its throat singing from a nearby branch. Off in the surrounding woods a yellow-billed cuckoo announces in dolorous tones that the rain will continue.

A bird dressed in raiment the color of Gainsborough's "Blue Boy" serenades me for several minutes with a sweet, simple song. It is an indigo bunting, one of the few birds that does more of its singing in July and August than in early spring. It is not a shy bird and can often be seen on telephone wires, tops of bushes, or even from the heights of tall trees.

Barn swallows are out in full force and making low-level flights over the insect-flooded field. They fly in erratic fashion, sometimes nearly hovering and other times moving in full-speed-ahead maneuvers.

I cannot see the bird, but from the hedgerow that separates the cornfield from the wide field, a yellow throat warbler pours out its distinctive song. I have no ear to differentiate such subtle musical notes, but this little fellow with black banded eyes is something special. It is always among the first to be heard in spring. By midsummer most listeners have become attuned to the plaintive song.

# Typical July

*The firefly wakens: waken thou with me.*

—Alfred Lord Tennyson

The sun went down last week, resembling nothing more than hell's fire on earth. All day it had moved in listless fashion across a dead sky, unrelieved by clouds that would break its fierce heat rays. Man and beast, fish and fowl moved to deeper shade. By noon a silence that could nearly be heard was in the air. It is a typical July day.

On the creek waterbugs and spiders skitter on the surface of the warm tidal flow that makes a slow passage to its headwaters. In the stillness of the air their paths are not silvered by errant breezes. Fish, made lazy by the heat, are in no mood to prey on tiny bugs.

About three hours after the sun sinks behind a cloudless horizon and shadows blanket the creek, I walk outside. A large misty moon has risen out of the east, and in its quicksilver light the marsh, creek, and woods start to come alive. A burnished ring around the moon with three stars inside the ring gives promise of changing weather in three days. This weather saying goes back many years and often is correct. (I remember only the times it comes true; the false alarms I conveniently forget.)

Lightning bugs are floating through the air by the thousands and emerging from the trees and undergrowth. The off-and-on radiance they spread is a thing of beauty. This lightning is caused by numerous chemicals in the bug's abdomen. When nerve stimulations release one chemical, the reaction causes light. When a different chemical is released, the light goes out. In some species the eggs and larvae also glow.

Lightning bugs, or "fireflies" to name them properly, are one-half inch and black with red or yellow markings. They are easily caught. In some countries natives confine them *en masse* in a jar for use as a lantern. Children, particularly those who live in the country, delight in capturing the tiny insects on a warm summer's night.

The larvae of lightning bugs live in the ground or in rotten wood or debris, where they subsist on small insects. According to some authorities, the adults feed on pollen, but other scientists think they do not feed at all.

As I continue my vigil, I become aware of a drumming of insects over the marsh, but, strange to say, not one was inshore where I stood. My mind flicked back hundreds of years to 1607, and I wondered how the early settlers of Jamestown were able to stand the constant harassment by hordes of insects that had to be prevalent then.

In the woods across the creek, a screech owl whimpers and in a few seconds, from far away, I hear another owl answer. They are wonderfully wise-looking little birds and when captured become quite tame. There are two color phases, brown and gray, but this is not determined by age, sex, or for an apparent reason. It just happens. The young birds are light gray puff balls of feathers, with large eyes, and lack the feathered ear tufts of the parents.

As the moon rises higher, it reflects tall cypresses and gums in the placid creek waters. Instead of trees, though, the shimmering light shows gargoyles, prehistoric animals, and unknown shapes from an unknown world. On the marsh, green pickerel weed is now the color of a silver blanket that robs every growing thing of an individuality. It is ghostly looking, and when I turn to go in the house, heat lightning in the western sky completes the illusion.

# Their Eden

*Oh, it's a snug little island!*
*A right little, tight little island.*

—Thomas Dibbin

If the heat and humidity of midsummer are getting to you, there is a simple solution to break the monotony. Take a boat trip to Tangier Island and let the breezes of Chesapeake Bay renew you to finish out the summer.

The tour boats are berthed at Reedville, Virginia, and two and a half hours should be allowed to drive there from lower Tidewater. The boat that my party travelled in was the *Captain Thomas*, seventy feet long, comfortable, safe, and capable of carrying 150 passengers. She was skippered by a young man from Reedville whose papers allow him to captain vessels up to 100,000 tons. The boat runs from May to September, and reservations should be made in advance.

The boats run out of Cockrell Creek as it begins the twenty-mile trip to Tangier. Each point along the creek shows evidence of existing or long-ago menhaden processing plants. Menhaden are found from Nova Scotia to South America and are usually captured in large schools that are located from the air. They average under twelve inches in length and are rich in oil that is extracted at the factory and mixed with cat and poultry food. The residue is used as fertilizer or a supplement to feed livestock.

Along the creek two families of nesting ospreys are seen. The nests were built on tops of buoys and under agreement with the Coast Guard would not be disturbed. The adults are handsome birds, dark brown above, and wearing a white baker's cap on top of their heads. Ospreys are only now making a strong comeback in the Chesapeake Bay area, after their numbers were decimated by indiscriminate use of pesticides and herbicides. They are strictly fish eaters, taking their food by spectacular dives into the water. Occasionally an oversized fish is hit, and the bird drowns, being unable to break loose the hooked talons sunk into its prey.

Cockrell Creek enters the Great Wicomico River, and here are some of the best oyster grounds in Virginia. Many pound nets are seen in the water, and around all of them are gulls, terns, and egrets trying to steal a meal. Also interesting is the Great Wicomico Light that used to be a manned lighthouse but now is fully automated.

From the Great Wicomico River our boat enters Chesapeake Bay with its five thousand mile shoreline. The skipper points out old Liberty ships of World War II vintage that make up a target range for flyers from Oceana Naval Air Station. Non-explosive missiles are used, but the old ships are pocked with many hits from low-flying planes.

Our boat arrives at famous Tangier Island at noon, and ten minutes later the entire group of visitors is seated at tables in the island's renowned eating establishment. Food is served country style, piping hot, and all of it is delicious. I am not a lover of seafood, strangely enough, but after the long water trip I eat crab cakes, clam fritters, and try a little of everything else that is passed by me. Our meal ends with a slice of golden yellow pound cake, light as an angel's wing and sweet as a baby's kiss.

Tangier Island is about five miles long and a mile and a half wide. Years ago it was much larger, but wind and waves have taken a toll of the land area. It has a population of nearly one thousand people, with most of the men making a living from the water. Crabbing is the principal occupation, and evidence of this is seen in the crab pots, traps, and shedding trays that dot the island. Oysters, fish, and clams are also caught, and in the wintertime, many waterfowl hunters come to Tangier seeking geese and ducks.

The island is not a place for everyone to live, and the residents are happy that this is so. They are independent people, clannish but friendly, and wake each morning with no clock to be punched. Theirs is an existence where each individual can do what he pretty well pleases. Crime is nonexistent, and religion plays an important part in their lives. Sometimes the weather is hostile when winter freezes come or hurricanes rip through to take more of the land area, but the Tangiermen and their families have learned to take it and live with it. May they continue to prosper and enjoy their Eden.

# Blackberry Hunt

*Blackberry stained lips,*
*July warm and purple.*

—Anonymous

The month of July has much to offer. With the exception of chiggers, ticks, and flies, all of it is good: moments like walking a beach, looking for prehistoric shark's teeth and arrowheads, or wading the shallows in a hunt for elusive soft crabs. Laughing gulls wheel in noisy fashion, accepting bread scraps from bikini-clad beauties. Green herons, handsome in their Robin Hood finery, move up from the beach in erratic flight.

Perhaps, though, you do not like the beach? If so, please join me for a fourth of July blackberry hunt. This is a traditional pastime that was initiated many years ago. I always look forward to it.

In order to avoid the stifling heat that usually accompanies July fourth, it is necessary to arise early, so six o'clock finds me eating a breakfast of hotcakes, eggs, sausage, and coffee so there will be no chance of starving before noon when we return. My hunting pants are tucked into my heavy socks, and hunting boots, a long-sleeved shirt, and straw hat are worn to protect against briars, snakes, chiggers, and other varmints that might be found in a blackberry patch. The straw hat is farmer's style, and inside its crown are two layers of water-soaked green leaves that are an outdoorsman's air conditioning. An additional protection against chiggers is the vile smelling kerosene oil that is liberally applied around ankles, waist, and arms. My final equipment is a wire-handled bucket that can be suspended from a belt, leaving both hands free for blackberry picking.

When we leave at six-thirty the sun is rising in a sky of turquoise blue. Not a cloud can be seen, and the atmosphere is so clear that it is unreal. It is already warm but not unpleasant. A fifteen minute ride takes us to the field where we are to pick, and a quick glance at the bushes indicates that no one has picked before us. We yell "Tally-ho" and plunge into the business of picking berries.

The bushes are loaded. Many berries are small due to so many being on the same branch, and not many of these are ripe. Others in the shade hang in clusters, like grapes. They are big, black, and shiny, and before many minutes pass will have to be sampled. Juicy and sweet, they are strong to the taste.

In the clear morning bobwhites are calling from several locations in the field. Barn swallows, flashing their orange breasts and cobalt blue backs, whip through the air only ten to fifteen feet above ground. Many insects will be dispatched by these busy birds, whose forked tails make them easy to identify.

The morning passes swiftly, and the berries rapidly fill the pails. Hands are stained with the royal purple that will require many washings to eradicate. Small scratches by briars are starting to sting, and a few imbedded thorns need to be removed with a needle. By ten o'clock, wasps and horseflies are out, and when one of the party sees a three-foot blacksnake lying in a bush, we decide to return home. Each of us has over two gallons of berries, and although we are now uncomfortable, everyone agrees it has been worthwhile. When we eat a blackberry cobbler some day this winter while snow covers the ground, this heat of the fourth of July will be a pleasant memory.

# A Well-Kept Lawn

*A child said, "What is the grass?"*
*Fetching it to me with full hands.*

—Walt Whitman

Many people take pride in a well-kept lawn. All year they will be found trimming, mowing (to a precise height), fertilizing, liming, watering, weeding, and generally manicuring a beautiful bit of greenery. The soil must be punched full of holes at intervals to serrate the grass roots, and woe betide the first unwary mole that plows a furrow through the yard. When this occurs, traps or poison are immediately brought forth to kill the little fellow that is merely trying to capture a grub or worm to eat.

If the grass does not do well, loads of topsoil are hauled in. This must be spread very, very smoothly and carefully raked to remove or break up clods of dirt, lumps, or debris. After this is done comes the introduction of a new grass seed, and there are many with which to experiment. Sometimes seed is not used and plugs of grass are inserted into the ground. Whichever kind is used, the true lover of beautiful lawns must eventually wind up with a plot of grass smooth as a baby's skin and shining like a newly polished emerald.

To these dedicated people, I take off my hat in salute. They have the patience of Job, the discipline of Patton, and the dedication of Sister Kenny. Also, they have a year-round job that brooks no interference from any other activity.

I also am an admirer of beautiful lawns, but only when they belong to someone else, leaving me free to fish, hunt, bird watch, and, in general, gallivant around the countryside. When I return home after such activities and am not too tired, then I can work on my lawn and know that its feelings have not been hurt by lack of attention.

In my situation, perhaps the word "lawn" is a misnomer. Better terminology is needed, so henceforth the word "yard" will be used. It is roughly two hundred feet long and perambulates around ill-kept flower beds, dogwood trees, garden spots (well-kept when weather permits), fig bushes, and grapevines. Nearly two hours of steady mowing is needed to cover it, and since other activities must not be curtailed, the grass is mowed every two weeks. In the late summer and fall, after the leaves commence to descend, I can fudge a bit and let it go for three weeks. My standard excuse then is that I must let the leaves accumulate so I can mulch leaves and cut grass at the same time.

The yard is completely surrounded by woods and is the only yard I know to be completely covered with vegetation without a seed being sown. In six years, only one hundred pounds of fertilizer has been used, and I think now this was a big mistake.

Recently I obtained a booklet describing 120 kinds of weeds and grasses of the southern United States. I am happy to say I have found over half of them in my yard. From early spring, when golden yellow dandelions dot the area, to late fall, when ragweed, goldenrod, and dog fennel try to escape the whirling blade of my mower, I am entranced by a wonderful array of plants that keep the dust down and give shelter to grubs, insects, and worms.

Honeysuckle with tubular blossoms edge their clinging tentacles from under dogwood trees and compete with stray bits of bright-leafed poison ivy for growing space. Virginia trumpet vine flaunts colorful blooms around the trunk of a pine tree and because of its beauty is not disturbed.

Henbit, crabgrass, wild bermuda grass (commonly called wire grass), buckhorn plantain (whose tapered flower spikes can be shot off by twisting the stem and pulling), all grow in profusion. Early spring finds large masses of white flowering chickweed that later are shoved aside and replaced by black medick. This low-growing plant is leafed like clover and has a tiny yellow flower in the shape of a tight cylinder. It is closely related to the forage plant alfalfa.

In season, buttercups and daisies appear. They are few, so most of the time the mower leaves them untouched. They make wonderful bouquets when picked and brought into the house by little girl grandchildren.

Other colorful grasses and weeds are scattered throughout the lawn. Broadleaf plantains with tall flower spikes, broadleaf dock, eastern bracken, common purslane, yellow woodsorrel, and wild garlic all add their bit to keep my yard covered. I enjoy them all; they require no care, and may they always come back.

# Great Scheme

*Every creature of God is good,*
*and nothing to be refused,*
*if it be received with thanksgiving.*

—The Bible

Records in the courthouse of James City County indicate that I own a certain number of acres on Powhatan Creek. Each year I receive a tax bill, and each year I pay my taxes. One would think that with this out of the way I would be sole owner and could control my acres. Don't think for a moment that this is true.

I began to think about this when the pestiferous phoebes tried to take over. I had no control over those birds, and they really took advantage of it. Suppose I did call the sheriff. When he came, could he catch and arrest the birds for trespassing or creating a nuisance? Would he think I was an idiot for calling? He would certainly be charitable if he did not.

How about the four raccoons that appeared at my door one night when I was working on the late shift? They were so indignant that no food was available that all of them climbed to the top of the back doorsteps and scratched loudly on the screen door, nearly scaring my wife out of her wits. To emphasize their protest, they wore masks over their eyes to prevent positive identification.

And the little bunnies—so cute and cuddly when seen in the yard— are bundles of brown fur who on summer nights become gluttonous, ravenous devourers of string beans, lettuce, and carrots. They seem to have an inborn radar that guides them away from the red-top clover and vetch that edge my garden and leads them to the delicacies inside. Is this the pound of flesh they demand because of my move into their area?

If it were not for their voracious appetites I would think them of the spirit world, for in winter and during hunting season they vanish at will. During one hunting season two friends came by with ten beagle hounds that "were the best in the world." Two hours of hunting by men and dogs failed to jump a rabbit. My friends left muttering to themselves "No rabbits," and I believe that, right then, all ten dogs could have been bought for two dollars. That night, two rabbits crossed my road in front of the car lights while a third was in my yard munching on frosted clover.

Foxes, coons, and 'possums are abundant, and they levy tribute on me like that of feudal lord. In five years, one rooster (a beautiful red one), two hens (good layers), two guineas (my security guards), and four mallard ducks have been taken in payment to keep my account balances. I am allowed to feed the remainder of my flocks until a fox decides her cubs are not getting a fair share of domestic meat or a mother raccoon thinks it high time for her young to feast on duck or chicken breast.

As landowner and taxpayer I would not have anything changed. The rabbits are a delight when they come out and play at twilight. The fox barking in the high woods and the coon squalling along the creek are sounds that are not given everyone to enjoy. The sonorous call of a hunting owl at night and the high-pitched scream of a hawk during the day bring the outdoors into the house. It is also pleasant to lie in a hammock and watch the swallows and swifts wheel overhead in their search for insects. A large turkey buzzard is the self-appointed department of health and does a good job of ridding the place of carrion.

The rest of the songbirds, snakes, turtles, and spring peepers are all part of the great scheme of nature. I am grateful that they share my land and allow me to support them.

# The Chickahominy

*Doubt not but angling will prove to be
so pleasant that it will prove to be,
like virtue, a reward to itself.*

—Izaak Walton

The sun was removing the blanket of night when we left the house. I had long promised by grandson a trip on the Chickahominy River, and when I called him at five o'clock in the morning, he responded with an alacrity not often seen in a twelve-year-old. In fact, a proposed fishing or hunting trip is the *only* thing that can get him up at that hour.

By six we are on the river and have our rods ready. From the end of each line dangles a favorite casting plug. As we move away from the dock, five wood ducks whistle by. In a group like this they almost certainly have to be two adults and three juveniles. Woodies usually hatch ten to fourteen eggs, but many little ducklings do not survive the perils of snakes, turtles, raccoons, and other predators.

In the marsh are male red-winged blackbirds, their scarlet epaulettes worn like a badge of courage, with their less spectacular mates enlivening the surroundings with early-morning chirping. Later in the day, we will hear the twinkling melody of male birds as they swing with abandon from the tall cord grass and shorter pickerel weed.

The marsh in August is decorated with huge rose mallows that, from a distance, resemble oversized helpings of vanilla or strawberry ice cream. They are lovely and grow in profusion in the slippery ooze of brackish, tidal marshes. Unlike many wildflowers, they are easy to transplant, and strange as it seems, do quite well in rich garden soil that receives no salt and only an average amount of water. The flowers in this area are the size of saucers and in the warmer part of the day furnish nectar to bees and other insects. Hummingbirds also frequent them, but are not evident this early in the morning.

The fish are not too cooperative, and I leave most of the fishing to my grandson. He responds by catching two small bass, both under twelve inches, that he promptly releases. Before returning them to the water, we admire the glistening bodies where bronze coloring is just becoming evident. Undershot jaws give them a pugnacious appearance that will increase with age.

As the tide carries us along, a large snake slides from the branch of a wax myrtle bush and disappears in the water. It is probably a harmless water snake, but could be a cottonmouth moccasin, a poisonous species occasionally found in local waters and swamps. It is about four feet long and as big around as a man's wrist.

By ten o'clock, barn swallows with forked tails and tree swallows with white-aproned breasts and bottle-green topcoats are numerous in the air around us. They are wonderful flyers, and occasionally we hear the click of beaks as they snap up unwary insects. Some skim the water low enough to drink while in flight, and one time a large fish breaks water in a valiant effort to take one for a meal. Reliable observers have reported seeing airborne swallows practicing this routine being caught by swift-moving fish.

A large raccoon wades the shallows in a daytime hunt for food. Its large, hunched back is characteristic of the bear family to which it belongs. It stops and takes a long look at us before scampering into the marsh. It will probably search for food until about noon, after which it will retire to some remote hollow tree to sleep until nightfall.

At noon, we head back to the landing. Only two fish are caught but the morning is memorable in numerous other ways, so we feel amply paid. Besides, we do not have any fish to clean, and this is always counted on the credit side. To paraphrase an old saying: "It's not how many fish are caught but how the game is played."

# A Recent Heat Wave

*O blithe newcomer! I have heard*
*I hear thee and rejoice.*
*O Cuckoo! I call thee bird*
*Or but a wandering voice.*

—William Wordsworth

Hot as the open door of a blast furnace, the sun scorched an earth already reeling from weeks without rain. Since early morn, rising from bed in the eastern sky, God's power plant of energy had again turned full blast on plants, trees, animals, and man. A relentless pressure, like a giant's hand, moved down and outward from a baby-blue sky that showed not a fleck of cloud.

Mature trees with roots driven deep in the ground seem the least affected by hot, dry weather. Massive root systems spread in all directions to suck moisture and food that sends life to widespread branches and leaves. The same roots serve as individual anchors, large and small, to hold trees against the onslaught of winds and storms.

Dogwood trees are shallow rooted and do not fare well in severe drought. Many branches show leaves half dry, and some have dead leaves. They await a strong wind to send them to their destiny with Mother Earth. Possibly some trees will make a comeback in the spring, but the fall will produce no ruby-tinted seeds to attract the many birds and mammals that feed on them.

Heat and lack of ample rainfall have a pronounced effect on the habits of other wildlife. Birds are most active in the early morning and late evening hours. Shady and swampy spots are favored during the middle of the day to conserve energy and strength. Desultory feeding on insects, spiders, caterpillars, and other minute forms of life takes place during these hours. Naturally, late-nesting birds with young to feed carry on their chores during the day. A pair of Carolina wrens nesting on my porch made countless feeding trips to care for four young birds during a recent heat wave.

Also excepted is the yellow-billed cuckoo whose guttural, repetitious call can be heard throughout the day. It is a sound that, like a rattlesnake's warning, once heard is never forgotten. Watermen, farmers, and country dwellers call the bird a rain crow because its call supposedly means rain is on the way.

The cuckoo is a slim, graceful bird, brown and white in color, and showing a distinct band of white across the tips of the tail feathers. A yellow lower mandible aids identification from its close relative, the black-billed cuckoo, which is seldom seen in Tidewater. The flight of both birds is strong and graceful, with much swooping and gliding.

64

Unlike the European cuckoos that deposit their eggs in the nests of other birds, the American species constructs a nest and raises its young. Shrubs and small trees are favored for nesting sites with the next choice being a flimsy platform of sticks and twigs. Eggs are usually two in number and are light blue in color with white cloud-like shadings. The nesting period is late May, June, and July.

The cuckoos have strange feeding habits. They prefer caterpillars, preferably the hairy ones like tent caterpillars, and will consume several hundred in a day's feeding. Other food consists of locusts, spiders, beetles, and flies. When a mayfly hatch takes place over a creek, cuckoos are always nearby to get their share. Little fruit and no grain is eaten so the silent flyers must be considered a definite asset to man.

In a search for moisture, four-legged animals adopt new methods. Rabbits move from thickets to chew on half-ripe cantaloupes, and actually cut down noxious pigweed and smelly marigold plants. Squirrels ruin entire corn patches and raccoons move out of the woods and marshes to take corn and melons. In a normal season with ample rainfall, damage occurs in garden spots, but not to such a great extent. Live traps will take many of these varmints, and they can be transported to areas distant from the garden.

# Mr. Bob White

*Now the day is over,*
*Night is drawing nigh;*
*Shadows of the evening*
*Steal across the sky.*

—Sabine Baring-Gould

It is late evening on the creek. All day—gray, gloomy, dismal—it has rained. Sodden clouds still hang over the fecund marsh of pickerel weeds that at a glance appear the same, but on closer inspection show no two alike—an inescapable fact of nature. No two things are perfect or exactly the same.

Thirty years ago the marsh mud of Tidewater was held together by heavy stands of cattails or reeds. In season, duck hunters could pull small boats into a marsh gut, throw out a few decoys, and hunt with perfect concealment from the tall growth. Today, marsh cattails and reeds have virtually disappeared and no one can answer with surety this question—WHY?

Some claim that various types of pollution have brought about the demise of such plants. Others say the marshes are settling and too much water is killing the root systems. I tend to subscribe to the second theory and would like to see on a trial basis several acres built up with dredge soil, and replanted with healthy root stock. If this attempt failed, perhaps another approach could be tried. A healthy marsh and the areas that depend on it need such plants.

At this time of day birds become more active. They seem to want to get in one more chirp or song before settling in for the night. A pair of gnatcatchers with elongated tail feathers chat in a nearby cypress tree. The white tail flashings and slim build make the little fellows easy to recognize at any time. In a nearby tree, a scarlet-coated cardinal whistles a cheery refrain that is answered by another bird across the creek. Friends at a distance, yes, but plumed fighting knights if either bird's territorial boundaries are violated.

In the distance, vociferous crows sound like they may be scolding an owl, many of which live in the woods that border the creek. Most owls when worried fluff up their feathers and sit like an immobile Buddha in an attempt to ignore the black tormentors. Seldom can they be provoked into flight since this makes them more vulnerable to attack.

Last spring, I placed a few bird houses along the edge of the creek to attract swallows. This year no swallows are using the boxes, but a pair of bluebirds moved in and are raising a family. I hope the immature birds wait until low tide to leave the nest. Otherwise, they will fall into the water where they will be easy prey to snakes, fish, or even a large bullfrog.

As I write, a pair of wood ducks squeal by with afterburners turned on. They have apparently been frightened and are really moving down the creek. They were close enough for me to see the crests and the white marking on the male bird's throat. Now a red-winged blackbird is warbling a night song from the branch of a dead cypress, and overhead, barn swallows in the buff and blue uniforms of colonial soldiers wheel and dart across the lowering sky. The fork-tail swallows make great inroads on the insect population and are welcomed wherever they appear. A word of caution: if you collect a barn swallow's nest, spray it first to kill the mites that infest it.

As night settles down a faint mist starts to fall. In the distance, Mr. Bob White, the little southern gentleman quail, whistles a lone good night. Shortly the lean, mean, creatures of the night will take over the creek and its marshes. Raccoon, mink, opossum, owls, and snakes will move with ferocious stealth to prey on careless creatures. It is the way of the wild—the strong survive, the weak perish.

# Summer Recipes

# Crab Soup

Mix two quarts crab meat, two quarts of hot water, one teaspoonful of red pepper and two of black, one large teaspoonful of salt, one bunch of parsley chopped fine, quarter of pound of butter. When it is well stewed, stir in one pint of cream or rich milk, and serve with crackers.

# Maryland Frog Legs

Boil the hind legs in salted water; squeeze on lemon juice from one lemon, then wipe them dry and dip in cracker crumbs and yolk of egg and milk; season with salt and pepper; fry them in butter and serve hot with French peas.

# Clam Fritters

Take two dozen clams, one egg, one cup milk, two small cups of flour or enough to make the batter; salt and pepper to taste. Chop the clams fine and stir them in batter made of the milk, and clam liquor beaten in the flour. Drip by the spoonful in boiling lard and fry brown.

# Shrimp Salad

Cut up the meat of freshly boiled shrimp. Squeeze over it some lemon juice with three tablespoonfuls of your best salad oil; let it stand on ice for one hour; then put it in salad bowl and pour over it a rich mayonnaise, and garnish with lettuce and hard boiled eggs sliced, and serve icy cold. Shrimp must be put alone into salted boiling water and allowed to cook for ten minutes; when they change color they are done.

# V-8 Juice

Wash and cut up one gallon ripe unpeeled tomatoes. Add three medium carrots (unpeeled) and three ribs of celery (cut up). Cook all of this until tomatoes have boiled to pieces. Strain off juice and bring the juice to boil again. Pour into hot sterilized jars. Add one teaspoonful salt and one teaspoonful Worchester Sauce to each jar. Seal and store.

## Chow-Chow

Take one peck green tomatoes, half dozen large onions, six green peppers (seed them); chop all together, and sprinkle with two or three cups of salt; let stand all night; in the morning squeeze out the tomatoes and throw away the water; put them all into a kettle with onions and peppers, a gallon of vinegar, one pound and a half of sugar, one teacup mixed black and white mustard seed, one tablespoonful cloves, one of allspice, two of celery seed; boil for twenty-five minutes.

## Deviled Crabs

Take meat from two dozen large crabs. Mix with one-half pound melted butter, one teacup of sweet oil, one-third teacup of cayenne pepper; one-half teaspoonful of black pepper, one tablespoonful of salt, one tablespoonful of mustard, one-half teacup of Worcestershire sauce, one-half tablespoonful of celery seed, four eggs, two handfuls of cracker crumbs; mix real well; fill the crab shells, sprinkle them with cracker dust, and bake a light brown.

## Grape Wine

Crush grapes and put in crock. Let stand five to eight days, stirring twice daily to immerse hulls in juice. Squeeze out juice. Add two to three quarts of water per gallon of juice. Add 2-½ to three pounds sugar per gallon and stir well. Put in jugs or cask, let stand three to four days. Water-seal until it stops working. Syphon off from dregs, bottle, and seal.

## Preserved Peaches (whole)

Pare the peaches and place them in cold water for a few minutes; allow one and a half pounds of white sugar to three pounds of fruit, and one teacup of water. When the syrup boils, put in the peaches, a few at a time, and cook until tender. Seal in glass jars.

## Blackberry Wine

Put one quart of cold water on twelve quarts of berries; press the juice and strain; allow three pounds of good brown sugar to each gallon of juice; put into a cask; leave uncorked until it stops working; put it in a cool, dark place, leave until December, then draw off and bottle for use.

## Watermelon-Rind Pickle

Boil ten pounds melon rind in water until tender; drain the water off; make syrup of two pounds sugar, one quart vinegar, half ounce cloves, one ounce ginger, and one ounce cinnamon; boil all this, and pour it hot over the melon.

## Old Virginia Mint Julep

Steep fresh mint in best brandy; make a syrup of sugar and water (use this sparingly); fill a tumbler (cut-glass) to one-third of brandy; put in three teaspoonfuls of syrup; fill the glass with pounded ice; stick five or six sprigs of mint on top and two or three strawberries. If old whiskey is used instead of brandy, more is required to make it strong enough. For an extra fillip, dash the julep with rum.

# FALL

# Old Autumn

*I saw old Autumn in the misty morn*
*Stand shadowless like silence, listening*
*To silence.*

—Thomas Hood

Fall comes to Tidewater as stealthily as a thief at night. The first sign on Powhatan Creek near Jamestown is the slight browning and yellowing of the marsh grass. Northeast winds encourage swollen tides that break the grass and hasten the drying process. In mid-September, hands clapped together bring forth high-pitched notes of protest from the elusive railbirds called sora.

These marsh birds are nine inches long, brown and gray in color, and have a short stumpy bill. They are similar to undersized bantam chickens and slip through the marsh with such celerity that they seem to literally walk on the water.

Years ago, a friend and I hunted sora in the Chickahominy marshes. We picked a day when a combination of full moon and northeast winds had produced the necessary high tide. Our boat was a shallow, double-ended skiff that could in theory be easily poled across and through the water-covered grass. It did for the first hour. After that, the eight-foot pole became an instrument of torture for seldom-used muscles. I pushed for ten minutes while my friend gunned, and then we reversed positions. The birds flushed just ahead of the boat and provided good but easy shooting. As I recall, the limit was twenty birds per hunter, and we collected close to our limits before the tide started out. It was a long trip back to the landing, where we cleaned the birds. Two nights later our wives served them up with wild rice, and they were delicious.

By September the black gum trees across the creek start their color change from dark green to russet. From then on, grackles, blue-jays, and other birds will feast on the blue-black berries that dot the branches in clusters. The true name of this tree is black tupelo, but it also is known as sour gum in the deep South and pepperidge in New England. It usually ranges from sixty to eighty feet high with a trunk three feet in diameter, but there are known instances of a height of 110 feet and a diameter of five feet. The wood is used to make boxes, baskets, and related objects.

The dogwoods are just a little later in changing from their summer frocks of green to a party-going red. Their limbs are festooned with scarlet berries that glisten in the fall sunlight and bear the seeds that will produce future trees.

The dogwood is the Virginia state flower, and legend tells us that its gnarled wood was used in the making of the cross that Jesus hung from. The edges of its flower petals are pink denoting the blood of Christ and the center represents the crown of thorns. The legend continues that because the tree was sorry for its part in the crucifixion, Christ ruled it would never grow large enough to be used for that purpose again. Virginia law now protects the tree from unnecessary cutting.

Dogwoods are easy to transplant and many could be saved if contractors would grant permission to people wishing to remove them from the path of bulldozers. If time is no object, sever the minor roots from around the trunk in December. Mark the tree with a piece of white cloth and leave until the last of February. By this time, the root system will be healed. Return, preferably on a damp or rainy day, and slide a shovel completely under the tree. This cuts the remaining root system. Leave as much dirt as possible, wrap in burlap, and transplant to a shallow hole lined with a mixture of peat moss and good soil. Tamp firmly and water. Mulch for two feet around the tree to retain moisture. If necessary, water at intervals until the tree takes hold.

# An Insect's Life

*The rules of the game are what*
*we call the laws of Nature.*

—Thomas H. Huxley

In the early days when the songbirds went on strike for a few days, the insects held a contest to determine the best singer amongst themselves. The stinging insects voted for the digger wasp, and the non-stinging ones went for the cicada, more commonly called the locust. The contest was even and not settled to anyone's satisfaction. It was on this day, many, many years ago that the digger wasp declared war on the cicada and all members of this family.

Recently, I saw a re-enactment of what has gone on through the ages, and the result was inevitable. The day was warm in September, and some time had passed without rainfall. Halfway up a gaunt cedar, a cicada was playing a loud, persistent song and was so pleased with its music that the swiftly flying wasp was on him before strong wings could be spread for flight.

The song died away in agony and the hot weather announcer lay still. He was not dead but paralyzed by swift-acting venom that the female wasp had injected into the nerve center.

The inert cicada was nearly twice the length of its foe, and I was interested to see how the wasp would transport its victim to a nest that had to be somewhere close by. My binoculars gave me a good view as she took hold with all six feet and then sailed through the air until the weight of the cicada brought her to the ground fifty feet away. With grim determination, the wasp slowly dragged the inanimate creature high up the trunk of a red oak. After resting for several minutes, she again launched herself into the air and landed at the edge of the clearing and near a pile of old bricks. I walked over and watched as she tugged and pulled her prey into a hole under the bricks.

I had other things to do, but curiosity got the best of me and I stayed to watch the finale. In about an hour the wasp reappeared and flew off. Moving some of the bricks disclosed a den that had been dug out about a foot into the ground and slanted down. At the very bottom was the cicada, still alive but helpless. On its body was one egg.

I covered the opened tomb with brick and a piece of plastic to keep it under observation. In three days the egg hatched, and a very small worm started to feed on the locust. In a week's time the cicada was completely devoured except for wings, and the little worm had grown to a fat white grub. In two weeks the grub spun a cocoon that will protect it through the fall and winter. Sometime next spring the cocoon casing will be split and a strong digger wasp will emerge to carry on its never-ending feud with the cicadas. Unlike birds, it will be on its own with no maternal care and yet, will function in the same way as its forebears of centuries gone.

There are many other types of wasps, and they are rated as being among the most intelligent insects on earth.

The two primary classes are social wasps and solitary wasps. Those belonging to the social group are the papermakers and it is said the Chinese followed their methods in the first manufacture of paper. The paper nests made by these wasps consist of wood pulp and plant fibers held together with liberal amounts of saliva. Larvae taken from these nests are excellent bait for bluegills, bream, bass, and crappie.

The solitary wasps are masons, carpenters, and excavators, with the most common being the familiar "mud daubers." Nests of these interesting insects are found under the eaves of houses or similarly protected places. To this group belongs the digger wasp.

Most of these insects are not aggressive and do not attack humans unless disturbed or by accident. The one exception is yellow jackets, and even they will not sting unless the nesting area is invaded. The best way to avoid getting "pinged" by this fearful group of insects is to learn a little more about them. Like some people, they require study, and after all habits are learned, they make excellent neighbors if left alone.

# The Wild Geese

*But when we hear the clarion honking
of Wild Geese, and, looking upward, see
the flying wedge clearing its way steadily
and rapidly through the sky, then we know
that we are seeing real bird travelers.*

—Frank M. Chapman

The wild geese come and explode the skies with the music of their coming. The wild sounds stop the farmer in his fields and the housewife at her work as they gaze skyward and watch the majestic birds. They are sounds that make one's hair stand on edge, as when "Dixie" is played by a band, or the skirling of a regiment of pipers fills the air.

Every fall, winter, and spring, they can be seen from Powhatan Creek, trading from the sanctuary at Hog Island to the vast rye and cornfields along the upper James. Powerful wings propel them through gusty skies like a sailing ship underway in dirty weather. Their honking can be heard for three miles, and in the safety of protected foraging the leader brings them to the ground with amazing speed. Once on earth, they gabble for a short time, and then some of them settle down to feed while the rest take up sentry duty. Never do they all put their heads down to feed at one time. If another flock heaves into sight, the sentries start talking them in and the feeders join in the chorus. After the newcomers land, a hubbub takes place that reminds one of a family reunion where relatives have not visited together in many years.

Geese are closely related to ducks and swans and are usually found in the same type of habitat. They are migratory birds ranging north to the Arctic Circle in the summer and as far south as Mexico in the winter.

There are thirteen kinds of geese in this country, with the best-known being Canada, snow and blue in that order. The dominant ones in Tidewater are the Canadians. They are excellent swimmers, and a wounded one can easily dodge a motorized boat. Heavy layers of down under their dense feathers insulate them against the cold waters that they frequent. Goose down was much sought after in the early days for comforters, quilts, and pillows. A gland located near the bird's tail produces oil that they rub on their feathers, rendering them waterproof.

The Canada goose is the largest of the group, ranging in length from thirty-six to forty-three inches. Its head and neck are black, with the exception of a prominent white patch extending from the upper side of the head, completely under the throat, to an equal distance on the opposite side of the head. The rest of the plumage is a brownish-gray. Its wing spread is five to six feet.

The food of this spectacular bird consists entirely of vegetable matter. This is taken as marshy roots, seeds, and berries, as well as grain from cut-over cornfields or the tender shoots of rye planted as a winter cover crop. There is a sand bar along the Jamestown Parkway that is a favorite spot for geese to pick up the gravel and sand they need for digestion.

Other names for the Canada goose are wild goose, black-headed goose, long-necked goose, and honker. Longfellow in his "Song of Hiawatha" gives the goose an Ojibwa name:

> And in flocks, the wild goose, Wawa
> Flying to the fen-lands northward,
> Whirring, wailing far above him.

Wild goose hunting has been a heady sport for many years, but despite the increased pressure by sportsmen, the big bird is holding its own. Ducks Unlimited, a program initiated by hunters, has played a large part by providing funds to purchase marshlands where *all* waterfowl are protected and can safely reproduce. It is probable that even with all this help, the geese will never be seen again as T. Gilbert Pearson reported when he saw a flock that took two hours to pass. That occurred in nearby Currituck Sound in North Carolina.

# Northeaster

*And the rain was upon the earth,*
*forty days and forty nights.*

—The Bible

Fall rode in on the tail of a typical Virginia northeaster. For three days it rained and rained and rained. Little puddles became big puddles, big puddles became ponds, ponds became lakes, and the lakes overflowed into marshes and swamps. It was so wet that even the frogs became mildewed.

Swollen tides absorbed the run-offs, and the creek became the color of coffee that has been laced with cream. Small logs, branches, and twigs, plus bottles and cans thrown overboard by thoughtless boaters, were swept along and lodged inshore when the waters receded.

Muskrats, many of them immature, were seen in the woods taking refuge from the high water. Herons were also feeding close to the bank, while kingfishers ranged up and down the creek, trying in vain to see small fish in the turgid, muddied water.

The garden area was soppy, slushy, sloppy, soaked, and saturated. Tomato vines that gave a bountiful yield only a few weeks ago were slumped against supporting stakes like worn-out dancers in frowzy costumes. A few pitifully small green tomatoes clung to weathered branches. Their destiny would be to ripen into red, mature fruit before Jack Frost rode through with his white paint brush.

The corn patch stood like a troop of Confederate soldiers, dressed in butternut-stained uniforms, bedraggled, bent, broken, but too proud to fall. Later on, a corn knife would slash them down even as Union sabers slashed down the stalks of the Confederacy.

During the third night the rain ceased. Next morning the sun rose in a world of wetness, cooled by a stiff breeze from the northwest. Fleets of sailing ships with full canvas unfurled sailed through a cerulean sky.

Tree swallows and chimney swifts were outlined against the backdrop of blue. They were high in the air, pirouetting through the sky like a troupe of aerial ballet dancers. High above them, a pair of red-tailed hawks made lazy circles. Their consummate skill enabled them to ride the airways with no visible movement of wings.

In the nearby woods black gum trees were the first to show a color change. Green leaves had turned to deep purple although many of the berries were still green. Later on they would ripen to blue-black and be eaten by birds and animals.

The marsh was full of colorful flowers, but many had been beaten down by the heavy tides. Cardinal flowers were still erect as were many of the yellow sneeze weeds. An occasional rose mallow was still in bloom but their place had been taken by the ruby-colored berries or rose hips of the wild rose.

In the garden, the row of zinnias had produced flowers, and many more would bloom before cold weather set in. Some of the blossoms had gone to seed, and this attracted American goldfinches to the yard. They were eye-catching birds, wearing dashing suits of brilliant yellow and black. I prefer the name that lots of people still use and from which I learned the bird—wild canary, or in some localities, thistlebird.

This small bird eats the seeds of various flowers but shows a preference for sunflower seeds. Like chickadees, it is not unusual to see them feeding head downward. In flight they are easily identified by undulating motion accompanied by intermittent chirping.

The cool weather had brought rosiness to the seed cones of the sumac bushes that abound along the wood's edge. As fall progresses they will turn deep purple, and by winter the dried seeds will be a welcomed addition to the diets of songbirds. To those interested in flower arranging, the wine-colored cones can be used with telling effect.

# Tidal Action

*I must go down to the seas again,*
*for the call of the running tide*
*Is a wild call and a clear call*
*that may not be denied.*

—John Masefield

Tidal action in waterways has always intrigued me, but like so many, I have accepted it and never bothered to go into the whys and wherefores.

A recent storm made me decide to find out a little more about this feature of nature and possibly help to enlighten others.

The heavy weather blew in ahead of a cold front. For several days, Indian summer had prevailed, with an approaching full moon lighting the night with a silvery glow. Shadows of leafless trees were outlined against the ground, and the image of a ghost owl could be plainly seen in its unhurried flight just overhead. A ring around the moon was in evidence for three nights preceding the storm.

An east wind ended the lovely weather and, together with the full moon, brought into the creek one of the highest tides that I've seen. First to be covered by the rising water were the brown pickerel weeds that infest the creek's edge. Next to be washed over were a few remaining cattail and wild oat stalks. Still the tide continued to march inland.

What impressed me most was the sight of a building floating slowly up the creek. At a distance it seemed to be fairly large, and when it finally lodged in a grove of silvery cypress trees on my side of the creek, I slipped on hip boots to investigate. However, the water was too high, and two days later, when I finally got to the building's resting place, I found it to be a well-built duck blind eight by ten feet. It had not been bushed for this season's hunting, so apparently it had been abandoned.

Tides are governed by both sun and moon, and one would think the gigantic size of the sun would exert the most control. Not so. Although the moon is smaller, it is also 390 times closer to earth and thus has a much greater influence on tidal action.

Tides rise and fall on schedule. Since this is so, the lives of many people are controlled by tide action. Fishermen, oystermen, and others who follow the water for their daily bread use tides more than they use a clock. High or low water means the difference in catching fish and also the difference in getting back to dock for the evening meal.

People who crab or clam for recreation know that low tide is the only time these can be taken, and surf fishermen seldom have luck on a low or ebb tide. Instead, they wait for the high waters that fill the sloughs with bait fish that are hunted by larger sport fish.

All bodies of water, salt and fresh, are subject to the forces of sun and moon. Lake Superior is one example. It has a tide range of about two inches, but like so many inland waters, the action is disguised by local winds and weather.

It has been known for thousands of years that the moon had a certain connection with tides. The Roman naturalist Pliny wrote of it, and the tidal laws were worked out after Sir Isaac Newton in the 1600s discovered the law of gravitation.

Normally, one tide acts like another. From low water, the tide flows inward for six hours. At that point—high tide—it recedes for six hours until low tide is reached. When the tide moves inward, it is called a flood current. On its outward flow, it is called an ebb current.

When the moon, sun and wind are all pulling along the same lines, abnormally high tides result, as was the case of the aforementioned storm.

Next day the sky was swept clean by a stiff northwest wind. In a short time the water had gone down low enough for familiar mud flats and bass-hiding logs to appear in the creek. By the second day everything was back to normal and will remain so until nature wills it otherwise.

# October Glories

*There is something in October
sets the gypsy blood astir.*

—William B. Corman

September has passed and the glories of October are on us. The main garden area has been mostly cleared up. Snap bean and lima bean bushes, heavily infested with Mexican bean beetles and their progeny, the small yellow larvae, have been pulled up and burned. Some bell pepper and tomato plants will produce until frost, and have therefore been left. Collards and three types of evergreen salads will provide good supply until the heavy freezes of January.

The secondary garden plot has a row of zinnias that still flowers profusely. Red, yellow, pink, and white blossoms vie with the gold and orange blooms of marigolds. Giant swallowtail butterflies resplendent in yellow and black costumes, black swallowtails with orange and blue dotted wings, and monarch butterflies, orange as a sunset, compete with honeybees, wasps, and bumblebees for life-sustaining nectar.

Two scuppernong grapevines are loaded with clusters of grapes, black as the night and sweet as ambrosia. It is impossible to walk by without picking a handful, which makes you want a second, third, or even fourth handful. A few late ripening figs and raspberries also aid in keeping one from famishing while walking around the yard.

The leaves of the dogwoods are richly colored and the branches are decorated with clusters of ruby-colored seeds. Squirrels and various birds are devouring them with a gusto that will clean the trees in a few days.

A few pokeberry bushes have been purposely left at the garden's boundary, and the ink-making fruit will hardly be noticed by birds until winter sets in. It will then be enjoyed by robins, thrashers, mockingbirds, and jays. The young shoots of poke are reputed to make good greens early in the spring, but someone else would have to verify this; I've never tried them.

Close by the zinnias, sweet potato plants cover the ground with massed leaf growth. The leaves are heart shaped, dark green, and have purple stems. The runners extend three to four feet on both sides of the row and sink their roots into the ground at intervals of two to six inches. The ground around the main plants has started to crack, indicating that good-sized potatoes have formed and will continue to grow. Around the end of October or sooner, if an early frost comes, they will be dug, dried in the sun for two weeks, and then stored. Later in the winter, baked in their jackets and broken open while steaming hot, their aroma will demand that I consume at least two big ones or three smaller ones.

As I write, a flock of eleven·domestic pigeons has flown in and landed on the dead branches of a white oak. Some are dark blue with pink throats, others are a lighter hue, and one has a light blue coat and white wings mottled with gray. They sit there very composed, cleaning their feathers and murmuring to each other with a guttural cooing sound. A bluejay flies up and scolds them a bit; he is ignored in a gentle way that soon discourages him.

Some years the oak trees are heavy with acorns, and many forms of wildlife gather and store them for use during the winter. The acorn crop is sparse this year, and a severe winter will cause hardship for the bird and animal kingdom. If this kind of weather comes along, we should put out as much food as possible to take up the slack.

At this time of the year the sumac is at its glory. The fern-like foliage ranges in color from a glossy dark green to burnt orange and scarlet. The wine-colored fruit is cone shaped and contains up to fifteen dwarf cones of the same color. It is often used in flower arrangements. During the winter, and after the seeds have turned brown, they are eaten by several kinds of birds.

The mature plant can be used as ground planting to cover the ugliness of banks or washes. It spreads rapidly, utilizing suckers as well as seeds.

# Many Pastimes

*Is not the gleaning of the grapes of Ephraim better than the vintage of Abiezer.*

—The Bible

Autumn offers many pastimes. Dove and rail hunting seasons have started, and the boom of shotguns can be heard from fields and marshes. Bow hunters are readying for the early deer season, and some counties in Virginia offer an early squirrel season. Migratory birds are moving through in flights south, and official "banding" stations are set up where birders trap and band them. I visited the Kiptopeke station on the Eastern Shore recently, and in two days over eight hundred birds were caught in mist nets, then given a numbered leg bracelet of aluminum and turned loose.

In fall, boaters, water skiers, and fishermen are having a final fling before winter sets in. On piers and creek banks crabbers offer chicken necks or tainted meat as bait to catch blue crabs that have escaped the fishermen. They are wary creatures but so greedy for the meat that they are easily caught with a long-handled net.

One of the things I enjoy most in autumn is searching for and gathering muscadines or "sloes"—the colloquial name given them by rural dwellers. Muscadines are wild grapes that grow throughout the South. In a favorable season, morning picking will usually yield several gallons. In many years the wild vines produce bumper crops. They are noted for sweetness and flavor and make a jelly that is hard to beat. When ripe the marble-sized grape is nearly black and separates easily from the stem. Clusters numbering up to a dozen may be found, but more often the musky grape grows singly or in pairs.

In the Old Testament is written: "And thou shalt not glean thy vineyard, neither shalt thou gather every grape of the vineyard." This admonition is easily followed when picking muscadines because the fattest and blackest grapes nearly always cling to vines so high in treetops that they are impossible to reach. Shaking the tree or vine will bring some fruit to the ground, but unless a tarpaulin or sheet is spread beneath, the grapes will be lost in the brush.

The grapevine is one of civilization's oldest cultivated plants. Grapes are mentioned in many stories, poems, and fables, one of the best-known being Aesop's "Fox and Grapes." Mae West also publicized the grape with her famous line, "Peel me a grape." Grapes grow in temperate zones throughout the world and are used in many ways. California produces 85 percent of the grape wine enjoyed in the

86

country. Special-type grapes are dried in the sun or by artificial heat and then called raisins.

Some of our popular domestic grapes like Concord, Worden, and Hartford were derived from the common muscadine. Scuppernong grapes that grow in the yards of many rural homes also were developed from the wild grape. The nectar from these is like pure sugar, and eating them is as addictive as eating peanuts. One calls for another.

Wild grapes are a favorite food of many species, among these being deer, raccoon, 'possum, skunk, fox, wild turkey, quail, pheasant, and songbirds.

To make an excellent jelly to use on winter morning pancakes try the following recipe:

Wash grapes thoroughly. Put in good-sized container that will not boil over. Bring to a slow boil and simmer until juice is completely out of grapes. Press through a strainer to eliminate skins, seeds, and most of the pulp. Reserve juice.

Using two cups of juice and two cups of sugar, bring to a slow boil and allow to boil for twenty minutes. Remove from stove, skim off foam, and pour into jelly glasses. After the mixture cools and hardens, cover top with paraffin wax.

If sugar is at a high price in the stores, the juice may be frozen and kept until prices come down.

# Otters

*Me thinks it is like a weasel.*

—William Shakespeare

The day was typical Indian summer. I had started early and by mid-morning was tired from following the meanderings of a tiny brook. A rugged obstacle course was created when tree laps from the lumbered woods had been allowed to remain as they fell. Many were lodged in the stream bed, and clambering through them, shod in hip boots, was not the easiest task I had ever undertaken.

I knew the brook eventually was confined by a beaver dam, but I had been interested in seeing if 'coon, 'possum, and mink had been feeding along the stream. A multitude of tracks and other signs confirmed my suspicions that many fur bearers fed along the creek each night.

When I reached a hill overlooking the beaver pond, I shucked off my camera, binoculars, and knapsack to take a few minutes' rest. A Joseph's coat of oak, beech, maple, and gum leaves made a comfortable seat that commanded a full view of the blackwater pond and its surroundings.

A pileated woodpecker, whose scarlet plume matched the handsome red patch on the head of a smaller downy woodpecker, took part in a drumming contest. The larger bird had the sound, but the gentleman in full evening dress of black and white was more rhythmic as it played its tattoo on a hollow beech log.

As I watched the birds, a splashing sound drew my attention to the picturesque body of water. I had anticipated the possibility of seeing otter signs, but I was pleasantly surprised to see three playboys of the weasel tribe cavorting in the water. They swam with a grace and agility not found in any other animal except members of the seal family.

Up and down the pond they raced. Sometimes the game was follow-the-leader and sometimes they did their own thing. One climbed out on the bank and belly-flopped noisily into the water, like a small boy going downhill on a sled. All the time they emitted a series of growling, squealing and chattering sounds that seemed to indicate what a fine time they were having. As suddenly as they had appeared, they vanished, and I wondered if I had dreamed of seeing them. The otters did not enter the scene again, and after a short time it occurred that they had probably winded me, causing them to vanish in some nearby retreat.

Otters are large mammals and are not nearly as rare as might seem. They are superb swimmers, and even the swiftest game fish like trout and bass have no chance against them. Like other members of the weasel family, an otter will often kill many more fish than it can possibly eat. When in this killing mood, usually only that section of the fishes' body just behind the head is eaten. Other meat they enjoy is that of muskrat, duck, young beaver, birds, and occasionally poultry.

The otter mates for life, and both parents help in rearing the one to five young that are normally born in April. The homes are established in comfortable burrows in a bank or hollow logs near water. The family stays together for at least a year, and during this time the young are given a full survival course.

Otters are strong animals, and can easily defeat a dog in battle. It has sharp teeth, short stout legs, and webbed toes. Under the two layers of fur is a blanket of fat that insulates against chill winter waters. Counting the sixteen-inch tail, they will often measure up to five feet, with a head as big as a medium size dog's.

Adult otters have a tendency to roam, especially when ponds and streambeds are frozen. On these travels they follow the smoothest route available but always remain fairly close to watercourses.

Otter pelts are considered one of the most beautiful and valuable furs in North America. They are used for fur coats and as trim for collars and cuffs. The fur is very durable, being graded 100 percent as compared with other furs.

Many otter are trapped each year, but with their wildness and habit of moving from place to place, it seems unlikely that trapping will ever remove more than the natural attritions that control all of our wildlife.

# The Hunt

*The huntsman winds his horn*
*And a-hunting we will go.*

—Henry Fielding

November marks the beginning of another hunting season. Since the first settlers at Jamestown took to the woods in search of food to see them through the winter, hunting has been a way of life for the men of Tidewater. Today, women have joined in the search for elusive deer, rabbits, and squirrels that are more numerous than in the early days of our country. Waterfowl is the only game that has shown a marked decline over those years but even so, enough ducks and geese move in from the north to ensure adequate shooting.

The abundance of game in Tidewater is due largely to the game commission. Staffed by trained professionals, the commission works toward a goal of more and better-managed game. Its wardens are respected and do an excellent job of enforcing game regulations despite the paucity of their number. A spring turkey gobbler season is a comparatively recent innovation that allows hunters additional time in the field. Large land tracts purchased by the state in Charles City and Surry counties provide places to hunt for the person who is not a member of a hunt club.

It would be nice if as pretty a picture could be presented on behalf of *some* hunters and so called sportsmen. Unfortunately, the actions of a minority can give the entire group a bad name. I am a hunter and have been for over fifty years and the following is not a sweeping indictment against all hunters. But there are careless people who take a gun and go in pursuit of game, and these "hunters" are nothing better than litterbugs, vandals, and gamehogs. Some are potential criminals—and there is no better name for them—who fire at sound or movement in the bushes without getting a glimpse of their targets. In the same category are those who mix alcohol and gunpowder while on a hunt. These are the people who give hunting a bad name and are responsible for the proliferation of "Posted" and "No Trespassing" signs that decorate the perimeters of woods, fields, and marshes.

I talked with an official of Chesapeake Corporation in West Point, Virginia. His words were: "You know, I almost hate to see hunting season come in again. Logging roads and fire trails through our property are torn up by four-wheel drive vehicles, signs are used for targets, and litter is increasing to an alarming degree." This, from a representative of a company that opens thousands of acres to hunters each year for a small permit fee.

Not long ago, three hunters occupied a restaurant booth close to where I was eating. One spoke in a loud voice about a dove hunt he participated in the previous week. The disturbing portion of the narration was: "Six of us killed two hundred doves last Saturday. We hunted from two p.m. until dark and I finally got tired of shooting." It is a shame that one of Tidewater's too few game wardens was not in the area to check the group on that day.

In days gone by, the poor and illiterate were generally blamed for poaching, night hunting, and other game violations. This is no longer true, if it ever was. Newspaper and magazine reports of today point to the pillars of the community as those most likely to violate the game and fish laws of the commonwealth. Their axiom seems to be "do as I say, not as I do."

At the same time, I am opposed to the anti-hunting segment of society that does not know and refuses to listen to, valid reasons about hunting as a sport, a recreation, and a means of proper game control. A case in point is Ducks Unlimited. No organization has done more for the conservation and protection of not only waterfowl, but other birds and mammals that inhabit the lands and waters it has obtained over the years from the Canadian government. I was therefore amazed when I attended a local DU supper to see three one-hundred-pound bags of corn auctioned off to raise money. This action implies tacit approval by Ducks Unlimited for baiting ducks and I cannot believe this to be true. Certainly the parent organization should disavow the practice through their fine magazine.

Most outdoorsmen are law-abiding citizens. They belong to outdoor organizations and work hard to establish good relations with farmers and landowners. They tag game properly and stop shooting when limits are reached. Severe weather finds them providing food to animals and birds that might otherwise starve. Educational programs are established for young people so they will know how to take care of themselves and others when taking to the outdoors. It is a pity that *all* hunters, fishermen, and outdoor people do not follow the same principles. If this could come about, everyone would benefit and the outdoors would be even greater than it is.

# Kingfisher

*I have laid aside business and gone a-fishing.*

—Izaak Walton

Even when ducks are not flying, a day spent in a duck blind can be rewarding. One reason is the presence of that never-tiring, belted knight of the waterways, the kingfisher.

The raucous battle cry of this piscatorial expert is heard from sunup to sundown in his never-ending pursuit of some unwary fish that ventures too close to the surface of the water.

The pattern is always the same. A diving splash that resounds over the quiet marsh and a triumphant return to a dead branch where the victim is swallowed head first. Immediately following is the hoarse rattle that announces to the world that another catch has been made and ingested.

Like a policeman, the kingfisher patrols certain sections of a marsh, lake, or pond, and guards it jealously. Most of the time, the bird flies from perches close to the water but occasionally he will hover, apparently having prey in sight, but still too deep for a successful catch.

The eyesight of a kingfisher can be compared with that of hawks and eagles. His dive is made from heights of fifteen to fifty feet and it is easy to imagine the difficulty in spotting a two- to three-inch fish from such a distance in water that can be anything from murky to laced with sunlight. The impact of a dive into the water should knock the wind out of the "fisherman" but he takes it in stride. Water temperatures also fail to deter the flying Izaak Walton who hits the icy waters of winter with the same nonchalance that he does in the balmy waters of summer.

Some human fishermen contend that their winged rival catches too many fish, but this charge has not been confirmed. Instead preferred food seems to be minnows, chubs, and other shallow-water fish. In hard times when fish are deep, the kingfisher will feed on crawfish, frogs, and some insects, especially those prevalent in watery surroundings.

The kingfisher is a picturesque but hardly handsome or pretty bird. His head, helmeted with a crest of gray feathers seems out of proportion with the rest of his body. Both sexes are chunky looking birds, grayish blue in color with a heavy band of blue across the chest of the male, and the female showing a rust-colored stripe. A sharp bill, longer than the bird's head, is used to spear its prey. The plumage is heavily oiled and with a suit of down underwear, the feathered "fisherman" is probably more comfortably attired than his human counterpart.

George Gladden, writing in *Birds of America* relates a legend of how the kingfisher received its not-so-common name of the Halcyon. The story is as follows:

> Alcyone, daughter of Aeolus, grieved so deeply for her husband, who had been shipwrecked, that she threw herself into the sea, and was immediately changed into a kingfisher, called "Halcyon" by the ancient Latin speaking people.

"Pliny says; 'Halcyons lay and sit about midwinter when daies be shortest; and the time whiles they are broodie is called the halcyon daies; for during that season, the sea is calm and navigable.'" The popular belief was that the seven days preceding the shortest day of the year were used in building the nest, and the seven days following were devoted to hatching the eggs. These fourteen days were called "halcyon days."

Today, the same adjective, halcyon, represents days of calm and peace when the fields, woods, and waters are at their best.

The kingfisher is one of the few birds that nest in the ground. The male and female join in the task of drilling a hole in the side of a clay or sandy bank, usually a tunnel nearly six feet long and six inches in diameter. At the end, a spacious nursery is excavated and here five to six white eggs are laid. The female is fed by the male while she goes about the task of hatching the eggs. When the young emerge, both parents take on the feeding duties. Most of the time, the young birds race to the front for food and then scurry back to await another trip by the parents. During the interlude, they maintain the usual quiet of young birds to avoid being taken by marauding mink, rats, snakes, or other predators.

# The Waterfowl Capital of the World

*"But hark! What sound is that approaching near?*
*'Down close!' The wild ducks come, and, darting down,*
*Throw up on every side the troubled wave,*
*Then gayly swim around with idle play."*
—Elisha J. Lewis

For four days in November, Easton, Maryland becomes the waterfowl capital of the world. The occasion is the annual Waterfowl Festival. I have visited there often and if the Good Lord is willing, I certainly intend to go there again.

The event is usually blessed with the Indian summer weather that is found only from Maryland southward. A few dark clouds may dot the blue sky and an invigorating breeze calls for only a lightweight jacket. In the parlance of the duck and goose hunting fraternity, it is "blue-bird weather."

The hub of the festival is the Tidewater Inn, a modern hotel of federal design. The personnel of the inn and volunteer workers who put on the festival are warm, friendly, and eager to assist visitors or guests. Leading out from the inn are sidewalks that have been decorated with pictures of flying geese. By following the painted figures, visitors are guided to the various points of interest.

The fee is nominal and permits entrance to all exhibits. An additional fee is charged to attend the auction or waterfowl-calling contest. Tickets can be puchased at many locations. Free shuttle buses provide transportation to all events. Comfortable viewing rooms are available where representatives of Ducks Unlimited and the Nature Conservancy show films of each organization's activities.

A Ducks Unlimited movie I watched dealt with the initial leasing of Canadian land by Ducks Unlimited, and the construction workers who built ponds, dikes, and culverts to control water for suitable waterfowl habitat, as well as hundreds of miles of fencing to protect nesting areas and miles of fire lanes to control marsh fires. Over a forty year period, Ducks Unlimited has spent $68 million, most of which has gone to create or stabilize close to three million acres of land that support not only waterfowl, but many songbirds and mammals.

The art exhibit alone is worth the price of admission. Artists from over twenty-five states and Canada display work that covers man's pursuit of the hard-flying game birds and other types of waterfowl. The paintings are so realistic that a spectator can almost sense marsh mud gripping hip boots, or feel the bite of sleet and snow driving out of the northeast. The artists know the habits of waterfowl. It is evident by the way decoying ducks or geese are painted coming into the wind. Their direction is indicated by blowing marsh grass.

The realism in the carving exhibit is such that people are heard to say, "How in the world can it be done?" Plumage on the waterfowl is so accurately etched that the birds look ready for plucking.

Many of the woodcarvings are truly outstanding, but one I saw that had most visitors standing in awe was that showing a fight between two red-tailed hawks and an eastern rattler that was making a valiant effort to escape. One hawk was a-wing with both sets of talons set in its prey and the other bird was lying on the ground with wings spread and one set of hooks in the snake. The snake was shown with mouth open and fangs extended trying to get in the fatal strike. The life and death struggle was evident in every detail and no one would guess that the creatures were of wood. The group carving was executed by Jules A. Bouillet of Vincennes, Indiana.

In the artifact display are decoys of all shapes, sizes, and descriptions. Many are primitive and have been collected through the years. Sneak boats, lanterns, and guns used by old-time market hunters are shown. Some guns are eight to ten feet long and have bores of two to three inches. They were built to mount on skiffs that were paddled into huge rafts of ducks. When the ducks erupted from the surface of the water, the small cannon was fired with disastrous results to the target. Most of this hunting was done at night and kills of one hundred ducks were not uncommon.

On Saturday an auction is held to sell decoys. Some are new and some are old but all are avidly sought after by collectors. Prices range from sixty dollars for an ordinary hand-carved duck to one thousand dollars and more for a Ward Brothers carving.

The festival is a huge success. Anyone interested in nature should make a special effort to attend it and gain an experience hard to equal. Congratulations to Easton, Maryland, for a fine job in teaching many people about our waterfowl heritage.

# A Month of Preparation

*Now in November nearer comes the sun
down the abandoned heavens.*

—David H. Lawrence

November is a month of preparation. Winter has not arrived, but the signs are evident. The red glow of a cold morning sun, which in Kipling's *Mandalay* "comes up like thunder," is enhanced by the silvery sparkle of white frost. Even redder sunsets outline wedges of flying geese in the western sky. Their ka-ronk, ka-ronk spills out of the sky as they move on the vast grain fields to feed and spend the night.

The pumpkins that filled the fields in October are gone but their memory is revived when one sights a persimmon tree, loaded with reddish-orange miniature fruits that resemble the Halloween symbol. Cold weather has given them a sweetness and taste not matched by anything else. Any night could find a big 'possum or 'coon high in the tree feasting on the fruit to build up fat for the winter.

Gray squirrels are busy as proverbial beavers, hunting acorns, hickory nuts, walnuts, and other goodies. They scamper about and when the time comes to bury or hide the prize, cast furtive looks around to be sure of privacy.

November is when blackbirds, grackles, and starlings flock together in masses numbering in the thousands. When they fly over, their wings create a whistling sound and often they will swoop down and over a chosen place many times before landing. At all times they fly as a unit, wheeling, diving, and circling. What sense enables them to perform these intricate maneuvers without crashing into each other?

By now the fur bearers have donned winter coats and by the end of November most will have a heavy layer of fat under the pelt to protect them against the rigors of winter. Some like the groundhog will hibernate until spring, but most will keep on the go except in extreme weather.

Birds that nested in the northern climes have returned. White-throated sparrows, towhees, and fox sparrows are seen in far greater numbers than during the summer months.

Man also has certain preparations to make in November. This is the season to collect wood for the stove and fireplace and in so doing, one is warmed twice—first, cutting the wood, and second, burning it in the fireplace. Without doubt, chainsaws make the task easier but crosscut saws and axes are not nearly as noisy and one gets a feeling of well-being by utilizing these tools.

There is a real skill in manipulating a long saw. The first requisite is a good sawing partner and the second is to make sure the saw is sharp. The secret to easy sawing is never to push on the saw. If each man pulls his load, the saw moves through a heavy log like a knife through hot butter.

Splitting wood with an axe can be hard work but a couple of tricks make it easier. Use a heavy axe, look for the minute cracks that are usually seen in the heart section of the log, and if the wood has knots, strike between them. Nearly any logs except dry gum can be split by observing these rules. For the large butt end logs, use a steel wedge and maul.

Early November is the time when hunters prepare for opening day. During spring and summer they have waited, and as the magic day approaches, they become restless. Rifles and shotguns are removed from cases and rubbed gently with oily rags that exude an odor all their own. Hunting clothes are retrieved from deep trunks or closets and put in a readily accessible spot. Quail or duck feathers may drift from a pocket and revive memories of a successful hunt. The clothes blend a fragrance of tobacco, woodsmoke, and marsh mud that cannot be appreciated by anyone other than a hunter.

November is the month that brings us Thanksgiving Day. It is a time to *stop* and *think* and to *give* thanks. We are fortunate to live in the greatest country in the world and our thanks should be extended to God, who has seen fit to shower many blessings on us. The world of nature was given by Him to us. Let us keep it in good shape and try to pass it down in a little better shape to those who follow us. I don't think we would regret it.

# *Fall Recipes*

## Roast Oysters (Maryland style)

Take large oysters, wash thoroughly and roast them over a hot fire. Three minutes time will find most of the shells opened and the oysters done. Finish opening over a deep dish so as not to lose any of the liquor; have butter melted; season with salt and pepper and place the oysters in it. Serve immediately.

## Jefferson Davis Punch

Mix one and a half pints lemon juice, three and a half pounds sugar disolved in water, twelve bottles claret, one and a half bottles sherry, half bottle brandy, one quarter bottle rum, one cup maraschino, three bottles ginger ale, six bottles Apollinaris or soda; garnish with two lemons, sliced thin, half cucumber sliced with peel, one orange sliced; if too strong, water may be added till the quantity reaches five gallons; best if made twenty-four hours before using, adding the ginger ale and Apollinaris just before serving. Serve with plenty of ice.

## Fruit Salad

Take Malaga or Concord grapes, oranges, and grated pineapple in equal quantities; soak in the juice of oranges and pineapple half box of gelatine. When it begins to harden, pour it over the fruit in layers; sweeten to taste and add candied cherries, and wine according to the quantity you prepare—half-pint of sherry to one quart of fruit.

## Collard Greens

Wash and trim tough stalks from collard leaves; barely cover with water; add one-half teaspoonful sugar; cook until tender; use knife and fork to cut into smaller pieces; fry about five or six slices of salted fat back until crisp but not burned; pour grease and meat over collards; after pouring off most of the water, simmer collards until ready to eat; salt and pepper to taste; collards are much better after being nipped by a heavy frost.

## Tenderloin of Beef and Mushrooms

Pare your filet nicely and season with salt and pepper; put it in a pan to cook for ten minutes, basting it constantly with butter; when done, take it out and cover with bread crumbs; make a sauce of Madeira wine, chopped mushrooms, a little tomato sauce; let cook for ten minutes and pour hot over the beef; garnish and serve.

## Virginia Brunswick Stew

Stew two large chickens until the meat leaves the bones; then chop it up and add one quart cut corn, one quart lima beans, three pints tomatoes, two onions, one quart snap beans, one quart okra; season with salt, pepper, Worcestershire sauce, two tablespoonfuls butter, and a little celery seed; boil all until it is well done and serve hot; for extra seasoning cut into small pieces two red peppers; squirrel can be used instead of chicken.

## Old Virginia Ham

Hams should always be soaked in cold water the night before cooking; then wipe it clean and place it in a boiler of cold water; let it be well-covered; simmer on the fire until done; it requires one hour to every two pounds of ham; when done, take out of water and remove the skin; pour grated bread crumbs over the top, little bits of butter and pepper, and set in an oven to brown; before sending to the table, stick cloves all over the top and garnish with parsley.

## Old Virginia Pork and Beans

Soak a quart of dried beans (preferably navy) overnight in warm water; change the water in the morning and put beans in cold water; let boil until soft; drain off the beans, and put them in a dish with two pounds of pork, which has been parboiled; add a little pepper and bake until light brown.

# Brandy Peaches

Use firm, clingstone peaches; peel them and lay them in cold water; to keep from turning dark, boil in clear water until tender; lay in dishes until cold; wipe thoroughly dry and pack in jars—a layer of peaches and a layer of sugar—until the jar is full; then pour on them the brandy; close the jars tightly; open near Christmas.

# Quail, Stuffed and Baked

Clean and wash birds thoroughly; be sure all shot and hidden feathers are removed; make a stuffing of bread crumbs, chopped celery, boiled chestnuts, butter, pepper, and salt; moisten all with a little hot water and stuff the birds; lay a thin slice of bacon on the top of each one; baste them while cooking with a little melted butter serve hot with tossed salad.

# Spice Cookies

To one pound butter add two pounds pulverized sugar, two pounds flour, one heaping teaspoonful each of cinnamon, cloves, and allspice; mix spice in flour with pinch of salt; roll out, cut small cakes, sticking a piece of citron in each; bake at once.

# Venison Haunch

Put the haunch in a baking pan; dredge with a little flour; have a gravy made with one pint of hot water, one pint of red wine, little cayenne pepper, two tablespoonfuls of butter, and a little salt; baste the venison with it and bake quickly; serve with currant jelly.

# WINTER

# Christmas Time

*Remember Christ our Savior was born on Christmas day.*

—Anonymous

Most poets do not treat the month of December kindly. In the poem "Snowbound," Whittier calls it a "brief December day;" Keats in "Stanzas" mentions a "drear nighted December;" and Edgar Allen Poe in his beautiful poem to the lovely Lenore distinctly remembers the "bleak December" when the Raven appeared. I tend to disagree with their descriptions of the month while reserving the right to thoroughly enjoy the same works.

December is the month chosen by Christians to celebrate the birth of Jesus Christ and certainly no more joyful event has occurred in the history of mankind. The beauty of Luke's presentation of the Holy Birth is unequaled by any other writer. The brightness can almost be seen as the angel gives the good news about the tiny baby lying in a manger in Bethlehem watched over by Mary and Joseph. The birth of this child was the spur needed to guide the footsteps of man along the right paths to a better civilization.

Christmas celebrations are varied, but nearly all countries utilize objects of nature in some form. In the Ukraine, a spider web decorates each Christmas tree in a traditional bid for good luck. Dutch children fill wooden shoes with food for St. Nick's horse, and an old English Christmas dinner always includes roast peacock and boar's head. Mince pies evolved from mutton pies and the pastries were originally cooked in the shape of a manger. Finnish villagers cut pine boughs and spread them in a green carpet for the Christ child from the top of a hill to the village center. Czechoslovakian girls break a cherry twig and place it in water on December fourth. If the twig blooms before Christmas Eve, she supposedly will marry before the year's end.

In countries where Christmas occurs during the warm season, flowers like poinsettia and the Noche-Bueno are used in lieu of evergreens.

The custom of using evergreens to decorate homes began in ancient times. The Romans exchanged green tree branches for good luck in January and the English are credited with moving the custom back to Christmastide. Holly, ivy, boxwood, and bay branches were used in those early days.

The Norse and Anglo-Saxons initiated the practice of burning the Juul (yool) to honor Thor, the God of Thunder. Later the Scandinavians changed the word to Yule, which meant Christmas. England next adopted the custom and it came to America with the early settlers. Today the Yule Log is burned every year in Williamsburg, Virginia.

Holly, mistletoe, ground pine, running cedar, pine cones, and magnolia leaves are widely used in this country for Christmas decorating. Most of these plants have a tendency to deteriorate quickly when placed in a heated house so they should be placed in water in a cool spot until ready for use. Holly is still plentiful but mistletoe, ground pine, and running cedar are definitely in danger and should be used sparingly. Ground pine can be cut so the roots remain in the ground but when running cedar is pulled, the root structure is destroyed.

# Audubon

*John James Audubon*
*Lived to look at birds.*

—Stephen Vincent Benet

In the Christmas story, according to the Gospel of Matthew, the Wise Men followed a star that appeared in the East and were guided by its brilliance to the birthplace of the Saviour. Luke's version describes the glory of the Lord shining over shepherds who were night-herding their flocks of sheep. Despite the difference in the writings, we know that spectacular events occurred on that night near Bethlehem.

Since that most important of happenings, other men have followed other stars. None have made the impact on humanity that comes close to equaling the birth of Christ, but many have made an infinitesimal contribution to the betterment of mankind.

Columbus followed a star that brought him to the shores of our country. The early settlers in Virginia, Massachusetts, and other colonies came after him and spread across the land to create a nation second to none. Many of them died for our freedom, but as they went down the long road to eternity, stars lighted the path they had taken.

John James Audubon was one who followed a star and he left us a legacy of bird knowledge and paintings that will endure forever. As a result of his influence, the birds of North America have been pictured and described with more detail than those in other parts of the world.

Audubon was born in Saint-Dominque, now known as Haiti. There is little doubt that he was an illegitimate child, but at a young age he was taken to France where he was formally adopted by his father, Jean Jacques Audubon. When eighteen years of age, the young man returned to America to work in earnest on subjects that had intrigued him since early childhood. He is best described in the first verse of Stephen Vincent Benet's poem:

> Some men live for warlike deeds
> some for women's words,
> John James Audubon
> Lived to look at birds.

At the time of Audubon's most prolific work, he was thrown in contact with the hodgepodge of people living between the Appalachians and the Rockies. Along the waterways, woods, and swamps he traveled, painted, and in general, lived the life of one who has a zest for adventure. He had a genius for not only being able to look, but to see. That is a real accomplishment and one found in few people.

From 1803 to 1849 he ranged from Florida to Labrador, from New York to Montana. He was equally at home with hardened frontiersmen like Daniel Boone or scholars like Daniel Webster. Audubon was not a loner. He appreciated good dress and had an eye for beautiful women or a pretty child. All men appealed to him—adventurers, crooks, gamblers, stuffed shirts, or the world's greats. His only request was to be accepted as a man among men.

The great painter and naturalist was not a businessman. Every effort at worldly trade that he undertook at various times failed. While he was in the woods and marshes painting birds, his wife, Lucy Bakewell, worked as a teacher to support the family. Again, the sixth verse of Benet's poem about him describes the situation perfectly.

> Followed grebe and meadowlark,
> Saw them swim and splash,
> (Lucy Bakewell Audubon)
> Somehow raised the cash.

In 1826, Audubon's star reached its zenith. He and his wife journeyed to England and Scotland where his pictures were rated as sensational. At this time he published *Birds of America*, a work consisting of 435 life-sized colored engravings made from his water colors. Despite his past record as a poor businessman, Audubon carried out the publishing venture, secured subscribers, collected from them, paid his workers cash, and made a complete sale of one thousand sets of one thousand dollars each. He returned in 1839 to the United States where he was the recipient of more honors. Death came to him at age sixty-one, but his work will live forever. Mankind is grateful for his contribution to our knowledge of birds and the world of nature.

# Duck Blind

*A duck blind on a wintry day is the*
*waterfowler's heaven on earth.*

—Bill Snyder

To some, a duck blind is a crude, bulky, inanimate, and probably downright ugly structure. The materials that make up a blind do not in most cases add to its beauty, and it is probably that these people view such a creation as a blot on the escutcheon of nature. Pursuing the matter a bit further, it is safe to say that lightning-blasted trees, awkward appearing animals or birds, and repulsive looking but beneficial insects are gazed upon with jaundiced eyes by the same people. Their argument seems to be "If it's not perfect, get rid of it."

I am a lazy individual who is unconcerned if a limb or a pine tree is bent at a grotesque angle or if leaves of some trees seem to persist in falling at all seasons of the year. The ugliness of a praying mantis or the awkwardly-hunched neck of a heron bothers me not in the least. In fact, they seem to have a character not found in more perfectly-formed species. There is no perfection in nature, so to myself and other duck hunters a duck blind is, in the words of the poet, "a beauty and joy forever."

Inspection of a newly built duck blind will in many instances tell as much about a man as a handwriting analyst or palm reader could point out.

Some blinds will carry the mark of a master craftsman, others will be constructed by your average duck hunter, and some will be thrown together (and this can almost be taken literally) by people like myself.

The blind of a true carpenter is put together with neatness and all the corners trued up to perfection. The cut cedar, pine boughs, or reeds will be symmetrical and woven through a wire netting with the precision of a master weaver's work. The floor will be sturdy and a comfortable seat is sure to be built in. Notches in which to lean guns for safety reasons and pegs to hold other necessities like duck calls, spare jackets, and extra weights for decoys will adorn the walls.

The average hunter and builder constructs a substantial blind, but one that will have knotholes or crevices that a whistling north wind will surely seek out. It is usually made from scrap lumber with burlap tacked to the sides. Over the burlap is placed any greenery available, or possibly dried reeds if enough can be found. Instead of wooden seats, a couple of five-gallon buckets scrounged from a nearby junkyard are available upon which to rest.

The last type of blind—and I must confess to using a similar one—is put together with anything available. A couple of dead limbs set in forked saplings constitute the cross bars. Other dead branches are driven into the mud and between all these are stuck branches and twigs of pine or myrtle bushes. More often than not, high tide finds me knee deep in water. This type of blind is cold, watery, muddy, soggy, slippery, and frankly, durned uncomfortable.

I still don't know why I duck hunt except for the fact that it is aesthetically satisfying. The journey to the blind in the false dawn and watching the world wake up at the real sunrise is an experience everyone should have. A lone hen mallard far away in the marsh quacks the morning reveille and she is answered from a distance by others of her kind. Overhead in the still dirty light, black ducks gabble in gurgling voices and the whistling wings of teal, pintails, and wood ducks sound like miniature jet planes.

In the early morning, sable-coated muskrats are often seen swimming the creek and the usual grumpy heron can be heard kronking across the marsh. Already a marsh hawk, white rump shining, is circling the marsh in a relentless hunt for unwary rodents or perhaps a crippled duck. A live duck in good health has nothing to fear from this handsome flyer.

Decoys bob on the water in front of the blind and suddenly are sighted by three swiftly-flying mallards. They make one circle and then come in as if pulled on a string. Two are drakes with bright green heads and the third is a hen in dull winter plumage. The gun speaks at one of the drakes and it crumples in midair. The two remaining ducks flare and grab with strong wings for more altitude. The hunter is satisfied and lets them go on their way. Even if no score had been made, the day would have been outstanding. Man, in the early morn of nature, is easy to keep happy.

# To My Grandchildren

*To be loved, be lovable.*

—Ovid

A new year is on us. The old one has passed leaving only memories. Some will linger but most will be, or have already been, shrouded in forgetfulness. Perhaps someday you will remember a small event that occurred in 1984 or maybe an important incident will accent that year in your minds. Whichever they are, may they bring happiness to you and those you love.

I have always found that the start of a new year is a good time to take a look at ourselves. The best way is not with a mirror that simply reflects our image, but to take a silent seat somewhere in God's world and think for a while. The place of your choice might be a lonely beach where dancing waves reverberate against the shore or a snow-drifted field where fieldlarks and quail or a stray rabbit glean the weeds for bits of food. A woodland, tall with evergreen pines and lofty oaks, or a sun-flected stream careening wildly to a distant sea are good places to rest and think.

I am sure at times that your young minds wonder and subconsciously rebel against the apparent gruffness and crustiness not only of me but also of your parents and other grandparents. Believe me, it has ever been so, and when you grow up to take our places, your littles ones will also wonder.

The reason we older people act the way we do to you is written in one word—LOVE. We want you to grow up to be responsible people in our society. We want you to accept the torch that we hand you and carry it on, and keep at least as good a world as you now live in. With your effort, the world can be made a much better place than we have bequeathed you. The words, "to you" are indicative of the discipline that perhaps at times you think is uncalled for or unnecessary. I can assure you that the same thing will be necessary when you take our places on the stage of life and have children of your own.

I want you to remember that God is first and foremost responsible for the beauty and inspiration given us in the world of nature. His is the breath that brings about the resurrection of tiny violets, arbutus, and other spring flowers that mantle the woods and fields. Nowhere is this better described than in a portion of the Old Testament poem, the "Song of Songs," attributed to Solomon:

110

For lo, the winter is past
The rain is over and gone;
The flowers appear in the earth;
The time of the singing of birds is come,
And the voice of the turtle is heard in the land.

He brings about the warmth of summer when the creation of all living things in the wild is kept up to date; He dresses the fall season in finery before everything is put to sleep and covered with the blanket of winter.

I want to ask all of you to make a sincere effort to follow some of the precepts that your parents and grandparents have tried to instill in you.

I am aware that all of you already love nature in one form or another. Because of this, you will not litter the woods, fields, and highways with cans, bottles, papers, chewing gum wrappers, and other unsightly objects. You will talk with your young friends and encourage them along the same lines. A chain reaction like this could have marvelous results.

As you grow older, you will become participants in peaceful movements to control pollution of our waterways, air, fields, and woods. United action of young people of your age can help bring back the clean world I knew as a boy.

Lastly, those of you who hunt will be sportsmen. You will take only game and fish to be used and you will obey fish and game regulations.

You will be careful with firearms and will teach others the same rules. Full bag limits will not be remembered nearly as long as that moment of ecstasy when a covey of quail bursts from under your feet or a brace of black ducks come to decoys after circling for twenty minutes. From all my days of duck hunting, I most remember an early morning when a friend and I paddled a skiff under cold stars whose light shimmered off frost-laden cattails. I hope some day this kind of memory will come to all of you.

Remember the words of Job, XII, 8, "Speak to the earth and it shall teach thee."

# Wild Wings

*Ducks, where do they hide when the season's in?*

—Bill Snyder

The last few weeks have seen an influx of ducks to the Chickahominy River marshes that makes me wonder where they all congre-
gated during the past hunting season. Nearly every year the same thing
happens. Very few ducks during the season, and two weeks later, the
marshes are full. Weather conditions are definitely not a factor.

I have talked with a number of duck hunters and nearly all have the
same opinion. They conclude that the ducks are heavily fed on the
sanctuaries during the open season and once legal shooting times are
past, the dispersal of food ceases and ducks move into outlying areas to
look for food.

I do not know if this is true, but if so, the duck hunters of Tidewater
are not being treated fairly. Most have invested heavily in decoys, boats,
motors, licenses, and other accoutrements necessary to pursue our
favorite sport. We stand behind Ducks Unlimited with full support,
knowing the fine job they are doing in buying up nesting areas where
not only ducks, but many other species of wildlife, will find protection to
raise their broods of young.

If the sportsmen's conclusions are true and the state is carrying out
a program of feeding ducks during the season to protect them, we hope
the practice will be discontinued at least two weeks prior to closing date
so hunters can enjoy good shooting for a short time.

For nearly two weeks, I have crossed the Chickahominy in the early
mornings and seen many ducks. Some days the weather was fair and
some days it was foul. Two days stand out in my mind and I'd like to take
you with me on those days.

The foggy day was outstanding. I slid the boat overboard at seven in
the morning and hooked on the motor. A lone rooster at a nearby farm
crowed the morning awake and a flock of domestic ducks gabbled an
accompaniment.

When I headed into the river, long, gray shrouds of fog enveloped
me. The murky wetness painted my foul weather gear so it glistened and
the same wetness gathered quickly on my eyeglasses. Visibility was only
fifty feet so it was necessary to hold boat speed to a minimum.

On the way, I passed within ten feet of a herring gull. It rode the
quiet water easily, feeding with intensity on a scrap of fish. Bright yellow
eyes flashed warily as the boat moved by, but the large bird rightly

assumed that I meant no harm so it continued to feed. At one time, the pearl gray and white plumage of this bird was much in demand for decorating ladies' hats, but this is now forbidden by law.

Several times I cut the motor and listened. Ducks by the hundreds were on the move so I assumed that the fog was hanging close to the water. In the distance, a flock of geese honked across the sky and the sound filtered through the murky air with amazing clarity. On this morning, much wildlife was heard but very little seen.

The second morning the weather was in direct contrast. Sunrise brought to mind an old German saying, "Morgenstunde not Gold in Munde" (The morning hour has gold in its mouth). Everything was clearly defined with not a cloud to be seen. A northwest wind chopped slivers of silver from the water and threw them helter-skelter into the air.

A red-winged blackbird, scarlet epaulettes shiny against a jet black uniform, swayed precariously on a slender reed and warbled a song of spring. Any bog, swamp, or marsh is brightened by the jolly, if repetitious, o-ka-lee, o-ka-lee of this bird.

Overhead, flocks of ducks were circling. Frightened by the boat, others winged up from the marsh and then settled back down as we passed. Pintails, the Beau Brummels of Ducktown, U.S.A., were numerous and easy to spot by needle-like tails.

Male mallards, flashily dressed as riverboat gamblers, flew so close that emerald green heads and the royal purple speculum were easy to see. They were followed by their coterie of females, whose petticoats were trimmed in lacy white.

Scaup or bluebills rode the choppy waters and were reluctant to move. When they did fly, all took wing at once and grey-white bodies were reflected by the morning sun. They are fast flyers and a joy to watch as they wheel and turn through the skyways.

# The First Snow

*Fire is the most tolerable third party.*

—Henry David Thoreau

The first snow of winter came in an unusual way. Normally, the flakes drift in under cover of nightfall, and when one looks out the window next morning, presto, the earth is covered with a blanket of white.

At six-thirty on the morning of this snow, I was crouched in a tide-flooded duck blind on a nearby creek. Rain beat down on my hooded foul-weather gear with the monotony of Chinese water torture; dark gray clouds scudding overhead indicated rain for the entire day. It was not particularly cold and my exertions in getting to the blind and putting out the decoys had warmed me to a perspiring state inside the rain gear. As I sipped a cup of coffee, I wondered why anyone in his right mind would leave a warm bed to participate in such an activity as duck hunting.

As day broke, bird life on the marsh commenced its daily activity. About the only kind not moving were ducks, and since this is not unusual when I hunt, I proceeded to enjoy the antics of the others. First to fly by was a kingfisher, so intent on fishing that I was not noticed. His eye was keen because shortly, a diving splash was made, and a triumphant rattle indicated the first catch of the day. A great blue heron went by on lumbering wings seeking a favorite fishing spot.

Some visitors came right into the blind. A nervy mite, gray in color with a tiny red crown, cocked an impertinent eye at me. It was a ruby-crowned kinglet and I appreciated the confidence it showed in coming close enough for me to see the identification patch. Cardinals flittered around in the nearby alder bushes. The conical seed cones of the water-loving shrub provides winter food for many birds. A myrtle warbler, bright yellow rump shining, searched industriously for minute insects, all the while keeping a wary eye on me.

The rain continued to come down in sheets, but I should have realized by the wildlife activity that a weather change was imminent. At ten-thirty, the water faucets of nature were shut off, the wind moved around to the northeast, and snow commenced. It drifted down silently, and shrouded the sedge-brown marsh with a ghostly whiteness. The leaf-covered surrounding woods were accorded the same treatment. All identity was lost to wind-driven flakes whose destiny sent them to draw a coverlet of winter over everything.

114

I stayed in the blind until noon. By then, cold had started to seep through my garments so I pulled the decoys and left. My destination was an old shack where pine lighter knots in an open fireplace would quickly restore heat to my body. The shack had doors and windows that are now open to the weather, but the roof was tight. In just a few minutes, turpentine-filled wood was ablaze and warmth edged into the room.

There is nothing quite like sitting by a crackling fire and watching snowflakes pile up everywhere they hit. The piney knots gave a smell all their own, the blue smoke danced a ballet up the chimney, and red-orange lights flickered off the logged walls. A few flakes falling down the chimney hissed their ways to oblivion in the fire. The sound, joined with the steaming of wet wood, was not unlike a tea kettle on the boil.

Emergency rations from the truck in the form of beans, franks, crackers, and cookies were broken out and devoured with gusto. The coffee was hot and steamy and warmed me all the way down to my toes. A beady-eyed mouse with mole-colored coat took bits of food that I threw on the floor.

The snow was wet and heavy and soon started to weigh the trees down. The branches of a cedar tree just outside the door kissed the ground and many of the pines were bent like a bow. Since I was a mile from the highway and on a muddy road, I decided to douse the fire with snow and leave.

Several times on the way out, it was necessary to stop, get out of the truck, shake snow from the trees bent over the road, and then go on. Eventually the highway came in view and from there, home was clear sailing. Score of the duck hunt? None killed but a very enjoyable day in the outdoors.

# Silent Flyers

The Wailing owl
Screams solitary to the mournful moon.

—Mallet—Excursion

The above quote exemplifies the sounds that emanate from the woods bordering Powhatan Creek on winter nights when moonbeams splash off silver-frosted cattails, and the incoming tide cracks and splinters ice that has formed in a recent cold snap. Instead of a solitary owl, there are sometimes as many as four breaking the night open with their wailing, hooting, screaming, and general caterwauling. The culprits, of course, are barred owls and it is difficult to realize they have such a repertoire of sounds. It is not so difficult to understand why people lost in the woods may be terrified at the unearthly owl music.

The barred owl has many *noms de plume*. Most common are hoot owl, swamp owl, rain owl, and round-headed owl. I coud give them other names—shrieking owl, arguing owl, bellowing owl, or dog owl. Some of their noise would make a banshee in labor sound like Brahm's "Lullaby."

They are beautiful birds with a coloration of pale buff and deep brown with equidistant bars on the upper parts of their bodies. The hooked bill is a dull yellow and the iris of their nocturnal eyes are dark brown with a nearly blue pupil.

Barred owls begin to nest as early as March. They usually select a hollow tree where two to four white eggs are laid. Occasionally they usurp the deserted nest of a hawk. When this occurs, sharp eyes can detect the non-sitting member of the family flattened out in a nearby fork of the tree.

The food of these nocturnal birds of prey consists of small animal life such as mice, rats, shrews, and an occasional rabbit, squirrel, or muskrat. Other food is taken in the form of frogs, lizards, crawfish, and large insects. Unless the prey is too large, it is swallowed whole, digested, and regurgitated later in the form of small pellets containing the hair and bones of the victims. Studies of these pellets have proven that barred owls are beneficial to mankind and should be afforded full protection.

The other large owl heard along the creek is *bubo Virginianus Virginianus* otherwise known as the great horned owl, Virginia owl, or cat owl. The feline comparison is well taken because when this great bird's wings carry it over marsh or woodlands, the noise is less than that of a cat stalking its prey, and its strike is deadly.

It is a rapacious, courageous, and powerful bird, and is not at all selective in its feeding habits. In unsettled areas, mammals of all kinds are taken, and in farming or poultry localities, domestic fowl are often killed. The lordly skunk is not even immune, often becoming an entree for the great horned owl. This is probably the only bird or animal that has the temerity to attack the scented gentlemen wearing the black and white coat.

The big owl lays its eggs in the latter days of February and in warm winters has been known to nest in January. Two to three white eggs are laid and when the young hatch, they resemble little puffballs of feathers with large eyes.

The adult bird is about two feet in length and has a wingspread of sixty inches. It is a handsome creature with dusty feathers mottled with an off-white coloration. The ear tufts are prominent and are nearly two inches long.

From studies made, it seems apparent that these birds have a capacity for much good or evil, depending in large part on their hunting grounds. In order to prevent the evil, it is wise to keep chickens, guineas, and turkeys locked up to protect them against this flying tiger of the night.

# Building a Fire

*Now stir the fire, and close the shutters fast,*
*Let fall the curtains, wheel the sofa round.*

—William Cowper

Today has been a drippy day on the creek. Windswept rain spawned from lowering gray-black clouds has saturated the earth from early morning. Now in the afternoon, the coldness has moved in and large blobs of marshmallow snowflakes fill the air. In the turbulent wind, they float, drift, and swirl before coming to a final resting place.

It is that kind of a day when it becomes mandatory to light off the open fireplace and relax in the warming glow of brilliant flames. I suppose everyone has a favorite method to build an indoor fire, but for those who don't, I will describe my way.

To begin with, and to stay on good terms with the rest of the household, be sure and open the damper. Otherwise, you might be ejected from a smoke-filled room to the chill outdoors where you can ruminate on your forgetfulness. Next, roll up several sheets of newspaper into tight balls and place at strategic points in the fireplace. Follow this step with a couple of handfuls of dry pine cones and criss-cross all of this with good-sized slivers of lightwood. Strike a match to the newspaper and in a few seconds, the entire mass will ignite and be ready for strips of dry kindling, preferably pine or cedar. After this, gradually increase the size of the logs and then sit back and enjoy the glass of muscadine wine made from wild grapes picked last fall. The inner glow combined with the outside warmth generated by the fire tends to put rose-colored spectacles on any individual.

For those born above the Mason-Dixon line, it is probably necessary to define "lighter," "lightning," or "lightwood." The name comes out of the deep south and probably is derived by the fact that such wood catches fire almost as fast as lightning strikes. Lightwood or "lighter knots" come from coniferous woods that have dried out and become heavily impregnated with pitch. Roots of pine that were cut many years ago are richly scented with a turpentine odor and give off tremendous heat when they start to burn. Most of these old stumps have been dug out, so it is now necessary to look for small short-needled pines that have died. After a few years of drying, these trees develop pitch, particularly in the area where the limbs attach to the trunk. Needless to say, permission should be obtained from landowners before taking such wood.

Lightwood burns with intensity and gives off a blue-black smoke that contrasts sharply against the deep burnt-orange flames. Some pieces crackle and pop almost in a happy fashion while others burn in a silent, vicious way. Split lightwood has a striking resemblance to a slab of bacon and in many sections of the South, it is called fatwood.

As darkness descends and snow continues, the fire takes on a magic not found elsewhere. It sends fingers of light stabbing into darkened corners where it reflects from aged furniture and pictures. From there, the fingers move over revered books that line several shelves.

Books by Zane Gray, Gene Stratton Porter, Robert Ruark, F. van Wyck Mason, and Stephen Vincent Benet are lighted along with nature books and duck-hunting volumes. *The Yearling* by Marjorie Rawlings has a permanent place as does *The Best Loved Poems of the American People* and *The Wise Fisherman's Encyclopedia*. The light seems to linger especially long on an ancient copy of *The Bible*.

There is little wonder that man has revered fire. It gave early man warmth and light to protect him from wild animals. In a short time, he learned to cook food, shape weapons, and make clay pottery. Today, fire funishes energy to drive machines that enable our vast industries to operate. It is used in the locomotion of mechanical equipment and it generates electricity.

Controlled fire is used to eliminate and destroy harmful bacteria and the wood industry uses this method to clear large areas of land for replanting.

Since fire was so hard to produce, many of our early civilizations kept a public fire which was never allowed to die. This practice was followed by the Egyptians, Persians, Greeks, and Romans.

So, the next time you start a fire, think a while about the ease with which you can accomplish this task. Be thankful that matches have replaced the flint and steel of the early Americans or the friction methods practiced by the Indians. We really have it nice, don't we?

# The Shortest Month

*It requires a strong constitution to
withstand repeated attacks of prosperity.*

—J. L. Basford

February is the shortest month of the year and not many people are sorry. Most days in February are damp, dreary, and dismal, or frosty, freezing, frigidly cold. Duck hunting season has gone and good fishing is still a month to six weeks away. It is the time of year when sneezing and runny noses are the rule rather than the exception.

The garden area is saturated and potatoes, peas, and other early vegetables cannot be planted. The usual winter birds like cardinals, white-throated sparrows, bluejays, and towhees have been seen so often that individuals can be recognized. They come to the wire-enclosed suet, or the yellow corn and wheat spilling out of the feeder, and look at it with jaundiced eyes as if to say, "Look, we've been feeding here all winter. Can't you come up with a wider variety of food in this dump of a restaurant?"

Perhaps February is to blame but at this point, I sit down and ask myself a question. Perhaps some others would like to cogitate on it. Are we, the people, doing birds and ourselves a favor by keeping them on a dole system throughout the year?

There is no doubt that a flock of cardinals, scarlet coats vivid against a backdrop of snow, presents a lovely picture. The picture is enhanced by the sergeant-major bluejays directing the traffic of foot soldiers made up of juncos, sparrows, starlings, and towhees. Chickadees, titmice, and woodpeckers of various kinds help to complete the kaleidoscope of winter's colors and make the entire scene a joy to behold.

In our selfish desire to enjoy these birds, have we forgotten that total welfare tends to weaken individuals? Is it not true that if continued long enough, free handouts are expected, and the desire for individual action or work is lost? A cardinal that does not have to forage for food is bound to have a weaker wing structure than those away from human habitation. Ground-scratching birds whose food is provided right on top of the ground have little need to exercise leg and foot muscles. What happens to these birds when a sharp-shinned hawk descends on them? Do they still have the exact amount of strength for a quick launch to possible safety, or have their underpinnings atrophied enough to make a difference? I don't pretend to know the answer but cold logic would seem to indicate that free feeding would exact a toll.

The sale of bird feed has increased tremendously in the last few years. It can be bought in just about any quantities, and despite inflationary prices, continues to sell. The manufacturers realize they have a good thing going and continue to introduce new mixes and feeders.

All creatures of nature have certain tasks to perform and that of birds is insect control. Masses of insect eggs, laid in the summer, will certainly not be cleaned up by birds that stop by a feeder regularly. The offspring of these eggs could very well attach themselves to your favorite shrub, tree, rosebush, or vegetable in the coming year. Instead of hammering away at a piece of suet, wouldn't it be a lot better for that downy or red-bellied woodpecker to perform the job assigned him?

This winter I have had food out for the birds. It has been a warm winter and very little food has been consumed. I intend to terminate the feeding at the end of February and will not initiate it next year except in case of severe snow or ice conditions. I plan to keep a close watch on the number of birds and what food they eat. My belief is that they will revert to natural foods like holly and sumac berries, alder cones, tulip poplar seeds, and weed seeds in general. I am sure the woodpeckers and all their traveling companions will scour the woods for insects and eggs that are hidden under and in the crevices of tree bark.

Perhaps it would be well to get back to the free enterprise system. Plant evergreen and berried trees or shrubs. They will provide other necessary items in their diets.

# A Storm in the Winter

When icicles hang on the wall,
And Dick the shepherd blows his hail
And Tom bear logs into the hall
And milk comes frozen home in pail.

—William Shakespeare

The later winter storm was born with a gentle northeast wind riding the airways. The red morning sunrise filtered through the gathering clouds and created a spectacular picture impossible for human hands to copy. The normally russet and tan marsh cattails and cord grass looked like they had been dipped in a paintpot of delicate rouge that changed the icy marsh into a fairyland. Despite the beauty of the morning, I was reminded of an ancient weather proverb: "Red sky in morning, sailors take warning," and I wondered if the weather in the next few hours would prove the validity of the admonition.

During the day, a small house barometer continued to drop and by nightfall thick, gray clouds blanketed the stars. A quietness was in the air and marsh and even the owls hunted like drifting ghosts. Late in the night I heard sleet slapping against the windows like bird shot and the rising wind moaning like a tormented soul.

When I arose at seven o'clock, the wind was whistling and driving the snow ahead of it, as if Mother Nature had emptied a huge pillow of white feathers. The flakes were large and all were in a vast hurry to rendezvous with a destination that would cover earth, trees, and buildings with a pristine mantle of white. All day it snowed and the cardinals, bluejays, and towhees wove multicolored contrails in their flights from protective bushes to the well-stocked larder of the bird feeder. Once during the day I counted fifteen splashes of crimson cardinals at one time, either feeding, resting, or in flight. Slate-colored juncos with pink beaks vied with the white-throat and field sparrows for morsels of food knocked to the ground by the larger birds.

Shortly after nightfall, I stepped outside and the cloud cover had moved, quietly retreating from the icy blasts of a now northwest wind.

Twinkling "forget-me-nots of the angels," a phrase used by Longfellow in his poem *Evengeline* to describe the beauty of the stars was certainly apropos for the winter night. There seemed to be millions of them, sparkling, shining and flashing their bright-eyed message of cheer to all the peoples of our world.

The day following the storm I spent walking on the marsh and in the woods. Squirrel tracks shaped like miniature snowshoes criss-crossed the woods in the four-inch snow. Before retreating to his leaf-lined hollow in a massive white oak, one little fellow took time to scold me in vociferous fashion, while bobbing tail kept time with his querulous barking.

A startled rabbit burst from beneath a snow-covered holly bush and with long hops moved through the woods, twisting and turning to be sure a bush or tree was kept between us. I lost the bunny from sight, but trailed it to a hideaway under a pile of dead branches where it again jumped. They are fun to watch in a situation like this, but come summer and garden time, when the same little rascal will cut down my snap-beans as fast as they break ground, he becomes a varmint.

Wading into the marsh showed me where a lone muskrat had come out of the creek to feed on cattail roots. Muskrat tracks are easy to identify, since in most instances the long furless tail is dragged in the snow or mud, making a separate mark between the footprints. Muskrats do not normally feed on a snow-covered marsh, preferring to stay in their lodges or burrows where they can feast on delicacies stored the previous fall.

A great blue heron flew from a protective nook and croaked a protest at being disturbed in his fishing. The long sharp beak and neck drawn back between wings reminded me of a hunchbacked Don Quixote with couched spear. It is not unusual to hear these birds at night. They apparently are disturbed by predators, and grumble along the creek until they find a sanctuary that will hide them until daybreak.

# The Varied Year

*See, Winter comes to rule the varied year,*
*Sullen and sad.*

—James Thomson

The world of nature has many contrasting moods: violence and serenity, beauty and ugliness, lightness and dreariness, happiness and sadness.

To some people, the period of sadness is autumn. Green grass under the violent stroke of Jack Frost has turned to drab brown. The same frost with a touch of ice has weakened the stems of leaves so they drift to the ground to die and decompose. Young birds have separated from their parents and are ready to join the exodus to warmer climates. The whole scheme of life seems ready to don a cloak of sadness.

I do not look on fall in this way. To me, it is a time of beauty, harvest, and fulfillment. The clamor of geese on an autumn day as they come home for the winter fills the sky. Wild ducks that were not here last week appear over the marsh in early mornings and late afternoons. In the fields, the boom of shotguns indicate that dove hunting season is in. The sound of the guns is interrupted by heavy machinery "bringing in the sheaves." Autumn is good.

The latter part of winter is something else again. If I have a low point, it has to be at this time. Let me explain why.

The gray, drab days of late February and early March are many. Rain-sodden clouds have deluged the earth until the soil is leaking at the seams. Boots are needed so one can walk in the yard and the squish-squash of rubber in mud is unpleasant to the ear. Puddles of water that in summer have a sparking clarity are brown and dingy looking and carry no invitation to children or birds to play in them.

The oak, poplar, and maple leaves that in the fall looked like a crazy quilt covering the ground are now a beaten, soggy mass. They lie inert and tight against the earth while a northeast wind vainly tries to dislodge them.

The dog shows the mood of the season. He stays in his shelter most of the time and sleeps the weather away. No cheerful tail-wagging or barking, and a squirrel skipping by evokes no interest at all.

At this time of the year, and particularly in bad weather, the birds lack the vivacity and cheerfulness shown at other times.

A flock of nearly fifty cedar waxwings chat in desultory fashion as they munch on holly berries that still dot the evergreen trees. They are sad little Quaker birds, quietly dressed and soft-spoken as they exchange the latest bird gossip.

Waxwings travel in flocks during the winter and their manner of flight is that of highly-trained flyers. They wheel, circle, and bend as if on command and suddenly drop with precision to holly, cedar, or cypress trees where they feed and rest.

Occasionally during the dreary days of later winter, a Canada goose, high in the void of gray clouds, sends out a haunting call. The sound is far different from that heard in the fall. It drifts down the airways like the faraway baying of a hound and the cry is that of a lonely bird searching for its mate.

By this time, large flocks of the noble birds have filed flight plans for nesting grounds in the far north. Sometimes in sinuous waves and other times in arrow-straight lines, they move through the skies like computer-controlled machines. The lonesome goose left behind is probably wounded or diseased, and senses the desertion of its companions and nesting partner.

During this season of the year, the creek winds through flats of mud instead of waterplants. What few stalks of tall marsh grass that dared grow in the polluted soil have been dashed over and destroyed by wind, tide, and rain. Water bugs do not skim the creek's surface and it is still too early for turtles and snakes to ease up on a mud-covered log to sun.

The one cheerful note that can always be counted on is the chattering and singing of the effervescent Carolina wren. Its song rings anytime during the day like a telephone attempting to awaken the world. When this sound is heard, the loneliness and sadness is pushed aside and the little brown bird with the broad white eye strip has eliminated such thoughts as expressed by Longfellow in "The Rainy Day:"

> Into each life some rain must fall
> some days must be dark and dreary.

# Unjustly Feared

*And there the snake throws her enamell'd skin.*

—William Shakespeare

The first warm days of later winter send me to the garden area. It is too early for general planting although potatoes and spring onions have been in the ground several weeks. Fencing has to be renewed to keep Br'er Rabbit out of the garden, and there are always the bean poles of last season that must be sorted out for another year's use. I was engaged in this not unpleasant task when I saw my first snake of the season.

It was a beautiful specimen of a pilot black snake, jet black on the back with light belly and white throat. It had apparently denned up for the winter under the stack of poles and was still in a half-dormant stage, hardly moving when I picked him up. The snake measured five feet eight inches long with a body diameter of two to three inches. The body was not distended, which indicated no food had been taken recently. After examination, the good-natured reptile was placed under some boards to resume its wait for warmer weather. At that time, it will commence cleaning up the unwanted rats, mice, insects, and possibly other snakes that inhabit the garden and woods area.

Snakes are interesting products of nature and probably are the most feared. A majority of people maintain vigorously that the only good snake is a dead one. This is a real tragedy since most of our snakes are beneficial to man and should be accorded protection. The only exceptions are the four venomous species: rattlesnakes, cottonmouth moccasins, copperheads, and coral. In a proper environment even they should not be molested.

A few facts may help to put snakes in a different light to those who encounter them occasionally.

Snakes are a part of the animal world known as reptiles. They are legless and deaf to sounds carried by air, but use their tongues to sense vibrations around them. Eyes are protected by transparent caps that never permit a snake's eyes to close. These caps shed when old skins are cast off, a molting process that occurs several times a year. Prior to this, the skin is a dingy gray and the eyes covered with an opaque screen. When the snake emerges from the old skin, it looks born again.

They are cold-blooded like all other animals, except mammals and birds. A snake's body takes on nearly the same temperature as its surroundings and this is one reason you never see a snake exposed to the direct rays of a hot summer sun for any length of time. A snake can stand much cold, and several times I have picked one out of a duck blind on a frigid winter morning, thrown it on the marsh like a dead stick, and watched it crawl away after the sun came up to create some warmth.

There are many legends and superstitions that surround these creatures. One says that snakes can milk cows. This is physically impossible because the rows of minute teeth in a snake's mouth would send a cow in a frenzy. Another concerns the hoop snake that rolls along, tail in mouth, until a victim is attacked. If the prey dodges and the tail hits a tree, the tree dies. Lastly, a rattlesnake's age can be told by the number of rattles. Not so. The skin is shed several times a year and a new rattle is added each time. Also, the rattles are brittle and tend to break off.

In most instances, there is little to fear concerning snakes. Tidewater has rattlesnakes, copperheads, and cottonmouth moccasins, but only in small quantities. The rest of the family are non-poisonous and generally useful, consuming many harmful rodents in a year's time. According to *World Book Encyclopedia*, scientists estimate that one snake will eat nearly 150 mice in six months. This would seem to indicate that snakes do belong and are a valuable asset in nature.

# A Good Day

*Oh! the snow, the beautiful snow,*
*Filling the sky and the earth below.*

—John Whittaker Watson

Today has been a good day on the creek. Last night a northeaster blew in and has continued today. The creek is pregnant with a heavy tide and rain is coming down in torrents. Great globs of storm clouds move inexorably across the sky and naked tree branches are festooned with globules of sparkling water—water that will become ice jewels if the temperature drops a few more degrees.

Our children and grandchildren are with us and have enjoyed the combination oak and coal fire in the fireplace. All of them have ohed and ahed to let grandpa know they appreciate the fire, and then have moved in to warm their backsides first, as generations of children and grandchildren before them have done.

The youngest daughter keeps asking, "Daddy, is it going to snow?"—as if I am a seer who always comes up with the wanted answers. "No," I reply, and am still child enough to hope that my answer is wrong and that tomorrow will see the ground covered with a cloak of ermine. The youngest girl grandchildren are in my lap and the oldest two boys are wanting to know how the squirrel and duck hunting is. Their daddy promises to take them hunting very soon.

Our oldest daughter is on cloud nine. Between mouthfuls of homemade succotash prepared by my wife, she proudly announces that she has finally had a magazine article accepted. It is a first after many, many rejection slips and we are as happy as she is. The article is about wild game cooking at which she is expert.

When they leave to return to their homes, I don foul weather gear and take a walk around the property. The wind is still strong out of the northeast, but the driven rain feels good against my face. I stop and inspect the hairy vetch in my garden and find it to be nearly two inches tall. It will not grow much more this winter, but will really take off in the spring. When it reaches a height of eight to ten inches, I will turn it under with a rototiller to put green manure into the soil. This will decompose and give off nitrogen that will ensure strong, sturdy garden plants in the coming spring.

As I walk, a flock of juncos, white-throated sparrows, and towhees fly ahead of me and into the surrounding bushes. The white-rimmed tail feathers of the juncos and towhees are not as showy as the patch of white feathers on the rump of a flicker that moves from the ground, where it has been feasting on ants, to a nearby tree stub. Like nearly all woodpeckers, this large member of the family is equipped with four toes, two forward and two aft, and a stiff-spined tail which enables it to brace with security on the side of a limb or tree trunk. It is unafraid and casts an inquisitive eye in my general direction.

On the way back, I find a cypress root where an otter has eaten a freshly caught fish. These strong swimmers are rarely seen on the creek but all wildlife leaves signs that can be read by any alert pair of eyes. Otters are large animals, very playful, and frequently cavort in the water like seals. They enjoy sliding down mud banks on their stomachs, a sign for which fur trappers are always on the lookout. A layer of fat covers the body under the skin and insulates the animal against the cold.

When I return to the house, the fire welcomes and cheers me. The flames bounce off the ceiling beams and pick up the varied colors in the stacked volumes of cherished books. In *Walden,* Thoreau says, "Books are the treasured wealth of the world and the fit inheritance of generations and nations." I can add nothing to this statement except "Amen."

# Winter Recipes

## Old Virginia Beaten Biscuits

Mix one quart of flour, one heaping teaspoonful of salt, two tablespoonfuls of butter and lard mixed, and enough milk to make a stiff dough (about one pint); work the dough a little, then beat with a flat-iron or biscuit beater until it blisters; roll out size you wish and cut with tin cutter; stick with a fork; bake at moderate temperature.

## Sally Lund Bread

Mix two and one-half pints of flour, three eggs beaten light, two ounces butter, two ounces lard, half-pint whole milk, three tablespoonfuls yeast, one teaspoonful of salt, one teaspoonful of sugar.

## Old Virginia Hoe-cake

Pour enough water on cornmeal to make it moist; add a little salt; put two tablespoonfuls on a hot griddle, which is well greased; make the cake half an inch thick; brown it on both sides; serve hot.

## For any punch made with alcohol:

A little water to make it weak,
A little sugar to make it sweet,
A little lemon to make it sour,
A little whiskey to give it power.

The proportion of water should ordinarily be two-thirds to one-third of the whiskey or rum, with a very little of the yellow rind of lemon and sugar to suit the individual taste.

## Old Virginia Sausage

Take five pounds of pork (fat and lean)—three times as much lean as fat, half pound pepper, two tablespoonfuls powdered sage, two ounces of salt; chop the meat and mix well with seasoning; keep in cool place; make into cakes and fry in hot lard.

# Apple Toddy

Take one gallon apple brandy, one quart of French brandy, one quart Jamaica rum. Bake one and a half dozen red apples; mash them, and pour on them one gallon and a half of boiling water; strain out the apples and sweeten the liquor with one pound and half of sugar; then add the brandy and rum and a few spices (cloves and allspice); pour in a demijohn and cork tightly; always make it some time before you need it; it will keep for years; roast several apples whole, and put them in bottom of bowl when it is served.

# Floating Island

Beat the whites of six eggs with any kind of fruit jelly or preserves that you like (strawberry or peach preserves are nice); add pulverized sugar to taste, and beat until *very stiff*; season a glass bowl of cream with a little wine or vanilla and sugar, and pile the island on top.

# Poor Man's Pudding

Take eight eggs, four tablespoonfuls of flour, four teacups of milk; add the flour to the yolks of the eggs; beat very light, and then add the whites beaten to a stiff froth; add the milk, and bake immediately for half an hour. To make sauce: mix one-quarter pound of butter, one egg, four tablespoonfuls of sugar, one wine glass of wine, well beaten together, and let it boil up only once.

# Chicken Broth

Take one chicken cut in small pieces; boil it until it falls to pieces in quart of water; then strain it and add two table-spoonfuls of rice or barley, which has been soaked in warm water; mix one cup of milk, salt and pepper, and a little chopped parsley or celery. Serve hot with toasted bread.

## Apple Float

Peel the apples and stew them until soft; press through colander, mash them very smooth and sweeten to taste; to one quart of apples, add the beaten whites of five eggs, and a little grated nutmeg; mix well and set on ice until wanted. Eat with rich cream.

## Egg-Nog

Take one dozen eggs beaten separately, one quart of French brandy, one pint of Jamaica rum; beat the yolks of eggs very light with one pound of sugar; then add the liquor a little at a time, stirring constantly; beat the whites very stiff, and mix nearly all with the yolks, leaving out a little to be put on top of mixture in bowl; lastly, add six pints of cream; grate a little nutmeg on top before serving.

## Macaroon Souffle

Take one pound macaroons mashed fine, one large lump butter, one pint hot milk, one tablespoonful of flour; melt the butter in saucepan, and then add flour; when well mixed, add the milk and macaroons; stir until it forms a smooth paste; then set this aside to cool; take a tablespoonful of butter, and stir in slowly the yolks of nine eggs, half-cup sugar, and the macaroon paste; lastly, beat the whites lightly; put all this in a baking dish well buttered, and bake one hour; serve with wine sauce, and stick blanched almonds over top.

RS '86

133

# About the Author

William Snyder was born in Newport News, Virginia in 1914 and grew up on a farm near what is now the Hidenwood section of the city. After graduating from Morrison High School in Warwick County, he began a forty year career in the Newport News Ship Yard. Since his retirement, Bill and his wife, Bridie, have lived in James City County near Williamsburg.

An outdoorsman all his life, he gardens and golfs, and enjoys duck and goose hunting, collecting old wooden decoys, reading and being a grandfather.

A prolific writer, his articles have appeared in *Virginia Wildlife* and *North Carolina Wildlife* magazines. He has written nature and personality columns for the *Virginia Gazette, Daily Press,* Gloucester-Matthews *Gazette-Journal* and *Tidewater Review.* His first book, *Wildlife Neighbors of the Williamsburg Area,* was published in 1981.